Important Physics Equations and Formulas

Physics is filled with equations and formulas. Here's a comprehensive list of some important ones to keep on hand — arranged by topic — so you don't have to go searching to find them. See the book's individual chapters for more!

Angular motion

- $\omega = \Delta\theta \div \Delta t$
- $\alpha = \Delta\omega \div \Delta t$
- $\theta = \omega_0(t_f - t_o) + \frac{1}{2}\alpha(t_f - t_o)^2$
- $\omega_f^2 - \omega_o^2 = 2\alpha\theta$
- $s = r\theta$
- $v = r\omega$
- $a = r\alpha$
- $a_c = v^2 \div r$
- $F_c = mv^2 \div r$

Electricity and magnetism

- $F = kq_1q_2 \div r^2$
- $E = F \div q$
- $W = qV$
- $C = \kappa\varepsilon_o A \div s$
- $E = \frac{1}{2}CV^2$
- $V = IR$
- $P = IV = V^2 \div R = I^2R$
- $B = F \div qvB \sin\theta$
- $F = qvB \sin\theta$
- $r = mv \div qB$
- $F = ILB \sin\theta$

Forces

- $\sum F = ma$
- $F_F = \mu F_N$

Gravity

- $F = G\, m_1m_2 \div r^2$

Magnetic field from a current loop

- $B = N\,(\mu_o I \div 2R)$

Magnetic field from a solenoid

- $B = \mu_o nI$
- $V_{rms} = I_{rms} Z$

Magnetic field from a wire

- $B = \mu_o I \div 2\pi r$

Mirrors and lenses

- $1 \div d_o + 1 \div d_i = 1 \div f$
- $m = -d_i \div d_o$

Moments of inertia

- Disk rotating around its center: $I = \frac{1}{2}mr^2$
- Hollow cylinder rotating around its center: $I = mr^2$
- Hollow sphere: $I = \frac{2}{3}mr^2$
- Hoop rotating around its center: $I = mr^2$
- Point mass rotating at radius r: $I = mr^2$
- Rectangle rotating around an axis along one edge: $I = \frac{1}{3}mr^2$
- Rectangle rotating around an axis parallel to one edge and passing through the center: $I = \frac{1}{12}mr^2$
- Rod rotating around an axis perpendicular to it and through its center: $I = \frac{1}{12}mr^2$
- Rod rotating around an axis perpendicular to it and through one end: $I = \frac{1}{3}mr^2$
- Solid cylinder: $I = \frac{1}{2}mr^2$
- $KE = \frac{1}{2}I\omega^2$
- $L = I\omega$
- $F = -kx$
- $T = 2\pi \div \omega$

Motion

- $v = \Delta x \div \Delta t = (x_f - x_o) \div (t_f - t_o)$
- $a = \Delta v \div \Delta t = (v_f - v_o) \div (t_f - t_o)$
- $s = v_0(t_f - t_o) + \frac{1}{2}a(t_f - t_o)^2$
- $v_f^2 - v_o^2 = 2as = 2a(x_f - x_o)$

Simple harmonic motion

- $x = A\cos\omega t$
- $v_x = -A\omega \sin\theta$
- $a = -A\omega^2\cos\theta$

Thermodynamics

- $C = \frac{5}{9}(F - 32)$
- $F = \frac{9}{5}(C + 32)$
- $K = C + 273.15$
- $Q = cm\,\Delta T$
- $Q = (kA\Delta Tt) \div L$
- $Q = e\sigma AtT^4$
- $PV = nRT$
- $KE_{avg} = \frac{3}{2}kT$

Work and energy

- $W = Fs\cos\theta$
- $p = mv$
- $KE = \frac{1}{2}mv^2$
- $\tau = Fr\sin\theta$
- $\sum\tau = I\alpha$
- $I = \sum mr^2$

Physics For Dummies®

Cheat Sheet

Conversion Factors between Measurement Systems

Here's a handy list to help you convert between systems of measurement to measure physical quantities. Keep it close for whenever measurement becomes an issue.

- 1 meter (m) = 100 centimeters (cm) = 1,000 millimeters (mm)
- 1 kilometer (km) = 1,000 m
- 1 kilogram (kg) = 1,000 grams (g)
- 1 Newton (N) = 10^5 dynes
- 1 Tesla (T) = 10^4 Gauss (G)
- 1 inch (in) = 2.54 cm
- 1 m = 39.37 in
- 1 mile = 5,280 feet (ft) = 1.609 km
- 1 angstrom (Å) = 10^{-10} m
- 1 slug (FPI system) = 14.59 kg

- 1 atomic mass unit (u) = 1.6605 x 10^{-27} kg
- 1 pound (lb) = 4.448 N
- 1 N = 10^5 dynes
- 1 N = 0.2248 lb
- 1 J = 10^7 ergs
- 1 J = 0.7376 ft-lb
- 1 British Thermal Unit (BTU) = 1,055 J
- 1 kilowatt-hour (kWh) = 3.600 x 10^6 J
- 1 electron volt (eV) = 1.602 x 10^{-19} J
- 1 horsepower (hp) = 550 ft-lb/seconds
- 1 watt (W) = 0.7376 ft-lb/s

Physics Constants

Physics constants are physical quantities with fixed numerical values, such as the speed of light or the mass of an electron. The following list contains the most common constants.

- **Avogadro's number:** N_A = 6.022 x 10^{23} mol^{-1}
- **Boltzmann's constant:** k = 1.380 x 10^{-23} J/K
- **Coulomb's constant:** k = 8.99 x 10^9 N-m^2/C^2
- **Electron's charge:** e = 1.602 x 10^{-19} C
- **Permeability of free space:** μ_0 = 4π x 10^{-7} T-m/A
- **Permittivity of free space:** ε_0 = 8.854 x 10^{-12} C^2/$(N-m)^2$

- **Mass of electron:** me = 9.109 x 10^{-31} kg
- **Mass of proton:** mp = 1.672 x 10^{-27} kg
- **Speed of light:** c = 2.997 x 10^8 m/s
- **Gravitational constant:** G = 6.672 x 10^{-11} N-m^2/kg^2
- **Gas constant:** R = 8.314 J/(mol-K)

For Dummies: Bestselling Book Series for Beginners

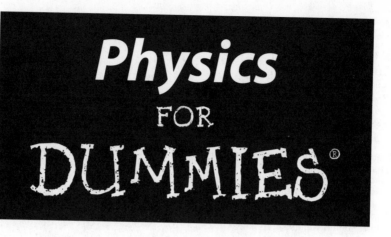

Physics FOR DUMMIES®

by Steven Holzner

WILEY

Wiley Publishing, Inc.

Physics For Dummies®

Published by
Wiley Publishing, Inc.
111 River St.
Hoboken, NJ 07030-5774
www.wiley.com

Copyright © 2006 by Wiley Publishing, Inc., Indianapolis, Indiana

Published by Wiley Publishing, Inc., Indianapolis, Indiana

Published simultaneously in Canada

For general information on our other products and services, please contact our Customer Care Department within the U.S. at 800-762-2974, outside the U.S. at 317-572-3993, or fax 317-572-4002.

For technical support, please visit www.wiley.com/techsupport.

Wiley also publishes its books in a variety of electronic formats. Some content that appears in print may not be available in electronic books.

Library of Congress Control Number: 2005933603

ISBN-13: 978-0-7645-5433-9

ISBN-10: 0-7645-5433-6

Manufactured in the United States of America

10 9 8 7 6 5 4 3 2 1

1B/RY/RR/QV/IN

WILEY

About the Author

Steven Holzner is an award-winning author of 94 books that have sold over two million copies and been translated into 18 languages. He served on the Physics faculty at Cornell University for more than a decade, teaching both Physics 101 and Physics 102. Dr. Holzner received his Ph.D. in physics from Cornell and performed his undergrad work at MIT, where he has also served as a faculty member.

Dedication

To Nancy.

Author's Acknowledgments

Any book such as this one is the work of many people besides the author. I'd like to thank my acquisitions editor, Stacy Kennedy, and everyone else who had a hand in the book's contents, including Natalie Harris, Josh Dials, Joe Breeden, et al. Thank you, everyone.

Publisher's Acknowledgments

We're proud of this book; please send us your comments through our Dummies online registration form located at www.dummies.com/register/.

Some of the people who helped bring this book to market include the following:

Acquisitions, Editorial, and Media Development

Project Editor: Natalie Faye Harris

Acquisitions Editor: Stacy Kennedy

Copy Editors: Josh Dials, Kristin DeMint

Technical Editor: Joseph L. Breeden

Editorial Manager: Michelle Hacker

Editorial Assistants: Hanna Scott, Nadine Bell, David Lutton

Cover Photos: © Getty Images/Photodisc

Cartoons: Rich Tennant (www.the5thwave.com)

Composition Services

Project Coordinator: Maridee Ennis

Layout and Graphics: Mary J. Gillot, Denny Hager, Erin Zeltner

Proofreaders: Laura Albert, Leeann Harney, Jessica Kramer, Arielle Mennelle, Joe Niesen, Carl William Pierce

Indexer: Joan Griffitts

Special Help
Danielle Voirol

Publishing and Editorial for Consumer Dummies

Diane Graves Steele, Vice President and Publisher, Consumer Dummies

Joyce Pepple, Acquisitions Director, Consumer Dummies

Kristin A. Cocks, Product Development Director, Consumer Dummies

Michael Spring, Vice President and Publisher, Travel

Kelly Regan, Editorial Director, Travel

Publishing for Technology Dummies

Andy Cummings, Vice President and Publisher, Dummies Technology/General User

Composition Services

Gerry Fahey, Vice President of Production Services

Debbie Stailey, Director of Composition Services

Contents at a Glance

Table of Contents

Part IV: Laying Down the Laws of Thermodynamics 205

Introduction

*P*hysics is what it's all about.

What *what's* all about?

Everything. That's the whole point. Physics is present in every action around you. And because physics has no limits, it gets into some tricky places, which means that it can be hard to follow. It can be even worse when you're reading some dense textbook that's hard to follow.

For most people who come into contact with physics, textbooks that land with 1,200-page whumps on desks are their only exposure to this amazingly rich and rewarding field. And what follows are weary struggles as the readers try to scale the awesome bulwarks of the massive tomes. Has no brave soul ever wanted to write a book on physics from the *reader's* point of view? Yes, one soul is up to the task, and here I come with such a book.

About This Book

Physics For Dummies is all about physics from *your* point of view. I've taught physics to many thousands of students at the university level, and from that experience, I know that most students share one common trait: confusion. As in, "I'm confused as to what I did to deserve such torture."

This book is different. Instead of writing it from the physicist's or professor's point of view, I write it from the reader's point of view. After thousands of one-on-one tutoring sessions, I know where the usual book presentation of this stuff starts to confuse people, and I've taken great care to jettison the top-down kinds of explanations. You don't survive one-on-one tutoring sessions for long unless you get to know what really makes sense to people — what they want to see from *their* points of view. In other words, I designed this book to be crammed full of the good stuff — and *only* the good stuff. You also discover unique ways of looking at problems that professors and teachers use to make figuring out the problems simple.

Conventions Used in This Book

Some books have a dozen conventions that you need to know before you can start. Not this one. All you need to know is that new terms appear in italics, like *this,* the first time I discuss them and that vectors — items that have both a magnitude and a direction — appear in bold in Chapter 4, like **this.**

What You're Not to Read

I provide two elements in this book that you don't have to read at all if you're not interested in the inner workings of physics — sidebars and paragraphs marked with a Technical Stuff icon.

Sidebars are there to give you a little more insight into what's going on with a particular topic. They give you a little more of the story, such as how some famous physicist did what he did or an unexpected real-life application of the point under discussion. You can skip these sidebars, if you like, without missing any essential physics.

The Technical Stuff material gives you technical insights into a topic, but you don't miss any information that you need to do a problem. Your guided tour of the world of physics won't suffer at all.

Foolish Assumptions

I assume that you have no knowledge of physics when you start to read this book. However, you should have some math prowess. In particular, you should know some algebra. You don't need to be an algebra pro, but you should know how to move items from one side of an equation to another and how to solve for values. Take a look at Chapter 2 if you want more information on this topic. You also need a little knowledge of trigonometry, but not much. Again, take a look at Chapter 2, where I review all the trig you need to know — a grasp of sines and cosines — in full.

How This Book Is Organized

The natural world is, well, *big.* And to handle it, physics breaks the world down into different parts. The following sections present the various parts you see in this book.

Part I: Putting Physics into Motion

Part I is where you usually start your physics journey, because describing motion — including acceleration, velocity, and displacement — isn't very difficult. You have only a few equations to deal with, and you can get them under your belt in no time at all. Examining motion is a great way to understand how physics works, both in measuring and predicting what's going on.

Part II: May the Forces of Physics Be with You

"For every action, there is an equal and opposite reaction." Ever heard that one? The law, and its accompanying implications, comes up in this part. Without forces, the motion of objects wouldn't change at all, which would make for a very boring world. Thanks to Sir Isaac Newton, physics is particularly good at explaining what happens when you apply forces.

Part III: Manifesting the Energy to Work

If you apply a force to an object, moving it around and making it go faster, what are you really doing? You're doing *work,* and that work becomes the *energy* of that object. Together, work and energy explain so much about the whirling world around us, which is why I dedicate Part III to these topics.

Part IV: Laying Down the Laws of Thermodynamics

What happens when you stick your finger in a candle flame and hold it there? You get a burned finger, that's what. And you complete an experiment in heat transfer, one of the topics you see in Part IV, a roundup of thermodynamics — the physics of heat and heat flow. You also see how heat-based engines work, how ice melts, and more.

Part V: Getting a Charge out of Electricity and Magnetism

Part V is where the zap! part of physics comes in. You see the ins and outs of electricity, all the way down to the component electrons that make action happen and all the way up to circuits with currents and voltages. Magnetism is a pretty attractive topic, too. When electricity flows, you see magnetism, and you get its story in Part V, including how magnetism and electricity form light.

Part VI: The Part of Tens

Parts of Tens are made up of fast-paced lists of 10 items each, and physics can put together lists like no other science can. You discover all kinds of amazing relativity topics here, such as time dilation and length contraction. And you see some far-out physics — everything from black holes and the Big Bang to wormholes in space and the smallest distance you can divide space into.

Icons Used in This Book

You come across some icons in this book that call attention to certain tidbits of information. Here's what the icons mean:

This icon marks information to remember, such as an application of a law of physics or a shortcut for a particularly juicy equation.

This icon means that the info is technical, insider stuff. You don't have to read it if you don't want to, but if you want to become a physics pro (and who doesn't?), take a look.

When you run across this icon, be prepared to find a little extra info designed to help you understand a topic better.

Where to Go from Here

You can leaf through this book; you don't have to read it from beginning to end. Like other *For Dummies* books, this one has been designed to let you skip around as you like. This is your book, and physics is your oyster. You can jump into Chapter 1, which is where all the action starts; you can head to Chapter 2 for a discussion on the necessary algebra and trig you should know; or you can jump in anywhere you like if you know exactly what topic you want to study.

Part I
Putting Physics into Motion

The 5th Wave   By Rich Tennant

After the circus, Bozo the Physicist went on to distinguish himself for his work on the Wave/Particle Joy Buzzer, squirting quarks, and Quantum Pratfall Theory.

In this part . . .

Part I is designed to give you an introduction to the ways of physics — also known as the ways of motion. Motion is all around you, and thankfully, it's one of the easiest topics in physics to work with. Physics excels at measuring stuff and making predictions, and with just a few equations, you can become a motion meister. The equations in this part show you how physics works in the world around you. Just plug in the numbers, and you can make calculations that astound your peers.

Chapter 1

Using Physics to Understand Your World

*P*hysics is the study of your world and the world and universe around you. You may think of physics as a burden — an obligation placed on you in school, mostly to be nasty — but it isn't like that. Physics is a study that you undertake naturally from the moment you open your eyes.

Nothing falls beyond the scope of physics; it's an all-encompassing science. You can study various aspects of the natural world, and, accordingly, you can study different fields in physics: the physics of objects in motion, of forces, of electricity, of magnetism, of what happens when you start going nearly as fast as the speed of light, and so on. You enjoy the study of all these topics and many more in this book.

Physics has been around as long as people have tried to make sense of their world. The word "physics" is derived from the Greek word "physika," which means "natural things."

What Physics Is All About

You can observe plenty going on around you all the time in the middle of your complex world. Leaves are waving, the sun is shining, the stars are twinkling, light bulbs are glowing, cars are moving, computer printers are printing, people are walking and riding bikes, streams are flowing, and so on. When you stop to examine these actions, your natural curiosity gives rise to endless questions:

- ✔ How can I see?

- ✔ Why am I hot?

- ✔ What's the air I breathe made up of?

- ✔ Why do I slip when I try to climb that snow bank?

- ✔ What are those stars all about? Or are they planets? Why do they seem to move?

- ✔ What's the nature of this speck of dust?

- ✔ Are there hidden worlds I can't see?

- ✔ What's light?

- ✔ Why do blankets make me warm?

- ✔ What's the nature of matter?

- ✔ What happens if I touch that high-tension line? (You know the answer to that one; as you can see, a little knowledge of physics can be a lifesaver.)

Physics is an inquiry into the world and the way it works, from the most basic (like coming to terms with the inertia of a dead car that you're trying to push) to the most exotic (like peering into the very tiniest of worlds inside the smallest of particles to try to make sense of the fundamental building blocks of matter). At root, physics is all about getting conscious about your world.

Observing Objects in Motion

Some of the most fundamental questions you may have about the world deal with objects in motion. Will that boulder rolling toward you slow down? How fast will you have to move to get out of its way? (Hang on just a moment while I get out my calculator . . .) Motion was one of the earliest explorations of physics, and physics has proved great at coming up with answers.

Part I of this book handles objects in motion — from balls to railroad cars and most objects in between. Motion is a fundamental fact of life, and one that most people already know a lot about. You put your foot on the accelerator, and the car takes off.

But there's more to the story. Describing motion and how it works is the first step in really understanding physics, which is all about observations and measurements and making mental and mathematical models based on those observations and measurements. This process is unfamiliar to most people, which is where this book comes in.

Studying motion is fine, but it's just the very beginning of the beginning. When you take a look around, you see that the motion of objects changes all the time. You see a motorcycle coming to a halt at the stop sign. You see a leaf falling and then stopping when it hits the ground, only to be picked up again by the wind. You see a pool ball hitting other balls in just the wrong way so that they all move without going where they should.

Motion changes all the time as the result of *force,* which is what Part II is all about. You may know the basics of force, but sometimes it takes an expert to really know what's going on in a measurable way. In other words, sometimes it takes a physicist like you.

Absorbing the Energy Around You

You don't have to look far to find your next piece of physics. You never do. As you exit your house in the morning, for example, you may hear a crash up the street. Two cars have collided at a high speed, and, locked together, they're sliding your way.

Thanks to physics (and, more specifically, Part III of this book), you can make the necessary measurements and predictions to know exactly how far you have to move to get out of the way. You know that it's going to take a lot to stop the cars. But a lot of *what?*

It helps to have the ideas of energy and momentum mastered at such a time. You use these ideas to describe the motion of objects with mass. The energy of motion is called *kinetic energy,* and when you accelerate a car from 0 to 60 miles per hour in 10 seconds, the car ends up with plenty of kinetic energy.

Where does the kinetic energy come from? Not from nowhere — if it did, you wouldn't have to worry about the price of gas. Using gas, the engine does *work* on the car to get it up to speed.

Or say, for example, that you don't have the luxury of an engine when you're moving a piano up the stairs of your new place. But there's always time for a little physics, so you whip out your calculator to calculate how much work you have to do to carry it up the six floors to your new apartment.

After you move up the stairs, your piano will have what's called *potential energy,* simply because you put in a lot of work against gravity to get the piano up those six floors.

Unfortunately, your roommate hates pianos and drops yours out the window. What happens next? The potential energy of the piano due to its height in a gravitational field is converted into kinetic energy, the energy of motion. It's an interesting process to watch, and you decide to calculate the final speed of the piano as it hits the street.

Next, you calculate the bill for the piano, hand it to your roommate, and go back downstairs to get your drum set.

Feeling Hot but Not Bothered

Heat and cold are parts of your everyday life, so, of course, physics is there with you in summer and winter. Ever take a look at the beads of condensation on a cold glass of water in a warm room? Water vapor in the air is being cooled when it touches the glass, and it condenses into liquid water. The water vapor passes thermal energy to the cold drink, which ends up getting warmer as a result.

Thermodynamics is what Part IV of this book is all about. Thermodynamics can tell you how much heat you're radiating away on a cold day, how many bags of ice you need to cool a lava pit, the temperature of the surface of the sun, and anything else that deals with heat energy.

You also discover that physics isn't limited to our planet. Why is space cold? It's empty, so how can it be cold? It isn't cold because you can measure its temperature as cold. In space, you radiate away heat, and very little heat radiates back to you. In a normal environment, you radiate heat to everything around you, and everything around you radiates heat back to you. But in space, your heat just radiates away, so you can freeze.

Radiating heat is just one of the three ways heat can be transferred. You can discover plenty more about the heat happening around you all the time, whether created by a heat source like the sun or by friction, through the topics in this book.

Playing with Charges and Magnets

After you master the visible world of objects hurtling around in motion, you can move on to the invisible world of work and energy. Part V offers more insight into the invisible world by dissecting what goes on with electricity and magnetism.

You can see both electricity and magnetism at work, but you can't see them directly. However, when you combine electricity and magnetism, you produce pure light — the very essence of being visible. How light works and how it gets bent in lenses and other materials comes up in Part V.

A great deal of physics involves taking apart the invisible world that surrounds you. Matter itself is made up of particles that carry electric charges, and an incredible number of these charges exist in all people.

When you get concentrations of charges, you get static electricity and such attention-commanding phenomena as lightning. When those charges move, on the other hand, you get normal, wall-socket-brand electricity and magnetism.

From lightning to light bulbs, electricity is part of physics, of course. In this book, you see not only that electricity can flow in circuits but also how it does so. You also come to an understanding of the ins and outs of resistors, capacitors, and inductors.

Preparing for the Wild, Wild Physics Coming Up

Even when you start with the most mundane topics in physics, you quickly get to the most exotic. In Part VI, you discover ten amazing insights into Einstein's Special Theory of Relativity and ten amazing physics facts.

Einstein is one of the most well-known heroes of physics, of course, and an iconic genius. He typifies the lone physics genius for many people, striking out into the universe of the unknown and bringing light to dark areas.

But what exactly did Einstein say? What does the famous $E = mc^2$ equation really mean? Does it really say that matter and energy are equivalent — that you can convert matter into energy and energy into matter? Yep, sure does.

That's a pretty wild physics fact, and it's one you may not think you'll come across in everyday life. But you do. To radiate as much light as it does, the sun converts about 4.79 million tons of matter into radiant energy *every second.*

And stranger things happen when matter starts moving near the speed of light, as predicted by your buddy Einstein.

"Watch that spaceship," you say as a rocket goes past at nearly the speed of light. "It appears compressed along its direction of travel — it's only half as long as it would be at rest."

"What spaceship?" your friends all ask. "It went by too fast for us to see anything."

"Time measured on that spaceship goes more slowly than time here on Earth, too. For us, it will take 200 years for the rocket to reach the nearest star. But for the rocket, it will take only 2 years."

"Are you making this up?" everyone asks.

Physics is all around you, in every commonplace action. But if you want to get wild, physics is the science to do it. This book finishes off with a roundup of some wild physics: the possibility of wormholes in space, for example, and how the gravitational pull of black holes is too strong for even light to escape. Enjoy!

Chapter 2

Understanding Physics Fundamentals

In This Chapter

▶ Understanding the concept of physics . . . and why it matters

▶ Mastering measurements (and keeping them straight as you solve equations)

▶ Accounting for significant digits and possible error

▶ Brushing up on basic algebra and trig concepts

*T*here you are, working away at a tough, nearly unanswerable physics problem, seeking a crucial breakthrough. The question is tough, and you know that legions of others have struggled with it fruitlessly. Suddenly, illumination strikes, and everything becomes clear.

"Of course," you say. "It's elementary. The ball will rise 9.8 meters into the air at its highest point."

Shown the correct solution to the problem, a grateful instructor awards you a nod. You modestly acknowledge the accolade and turn to the next problem. Not bad.

With physics, the glory awaits you, but you have some hard work waiting for you, too. Don't worry about the work; the satisfaction of success is worth it. And when you finish this book, you'll be a physics pro, plowing through formerly difficult problems left and right like nobody's business.

This chapter starts your adventure by covering some basic skills you need for the coming chapters. I cover measurements and scientific notation, give you a refresher on basic algebra and trigonometry, and show you which digits in a number to pay attention to — and which ones to ignore. Continue on to build a physics foundation, solid and unshakeable, that you can rely on throughout this book.

Don't Be Scared, It's Only Physics

Many people are a little on edge when they think about physics. It's easy to feel intimidated by the subject, thinking it seems like some foreign high-brow topic that pulls numbers and rules out of thin air. But the truth is that physics exists to help you make sense of the world. It's a human adventure, undertaken on behalf of everyone, into the way the world works.

Although the contrary may seem true, there's no real mystery about the goals and techniques of physics; physics is simply about *modeling* the world. The whole idea behind it is to create mental models to describe how the world works: how blocks slide down ramps, how stars form and shine, how black holes trap light so it can't escape, what happens when cars collide, and so on. When these models are first created, they often have little to do with numbers; they just cover the gist of the situation. For example, a star is made up of this layer and then that layer, and as a result, this reaction takes place, followed by that one. And — pow — you have a star.

As time goes on, those models start getting numeric, which is where physics students sometimes start having problems. Physics class would be a cinch if you could simply say, "That cart is going to roll down that hill, and as it gets toward the bottom, it's going to roll faster and faster." But the story is more involved than that — not only can you say that the cart is going to go faster, but in exerting your mastery over the physical world, you can also say how much faster it will go.

The gist of physics is this: You start by making an observation, you create a model to simulate that situation, and then you add some math to fill it out — and voilà! You have the power to predict what will happen in the real world. All this math exists to help you feel more at home in the physical world and to help you see what happens and why, not to alienate you from your surroundings.

Be a genius: Don't focus on the math

Richard Feynman was a famous Nobel Prize winner in physics who had a reputation during the 1950s and '60s as an amazing genius. He later explained his method: He attached the problem at hand to a real-life scenario, creating a mental image, while others got caught in the math. When someone would show him a long derivation that had gone wrong, for example, he'd think of some physical phenomenon that the derivation was supposed to explain. As he followed along, he'd get to the point where he suddenly realized the derivation no longer matched what happened in the real world, and he'd say, "No, that's the problem." He was always right, which mystified people who, awestruck, took him for a supergenius. Want to be a supergenius? Do the same thing: Don't let the math scare you.

Always keep in mind that the real world comes first and the math comes later. When you face a physics problem, make sure you don't get lost in the math; keep a global perspective about what's going on in the problem, because doing so will help you stay in control. After teaching physics to college students for many years, I'm very familiar with one of the biggest problems they face — getting lost in, and being intimidated by, the math.

And now, to address that nagging question plaguing your mind: What are you going to get out of physics? If you want to pursue a career in physics or in an allied field such as engineering, the answer is clear — you'll need this knowledge on an everyday basis. But even if you're not planning to embark on a physics-related career, you can get a lot out of studying the subject. You can apply much of what you discover in an introductory physics course to real life. But far more important than the application of physics are the problem-solving skills it arms you with for approaching any kind of problem — physics problems train you to stand back, consider your options for attacking the issue, select your method, and then solve the problem in the easiest way possible.

Measuring the World Around You and Making Predictions

Physics excels at measuring and predicting the physical world — after all, that's why it exists. Measuring is the starting point — part of observing the world so you can then model and predict it. You have several different measuring sticks at your disposal: some for length, some for weight, some for time, and so on. Mastering those measurements is part of mastering physics.

To keep like measurements together, physicists and mathematicians have grouped them into *measurement systems.* The most common measurement systems you see in physics are the centimeter-gram-second (CGS) and meter-kilogram-second (MKS) systems, together called *SI* (short for Système International d'Unités), but you may also come across the foot-pound-inch (FPI) system. For reference, Table 2-1 shows you the primary units of measurement in the CGS system. (Don't bother memorizing the ones you're not familiar with now; come back to them later as needed.)

Table 2-1	Units of Measurement in the CGS System	
Measurement	*Unit*	*Abbreviation*
Length	centimeter	cm
Mass	gram	g
Time	second	s

(continued)

Table 2-1 (continued)

Measurement	Unit	Abbreviation
Force	dyne	dyne
Energy	erg	erg
Pressure	barye	ba
Electric current	biot	Bi
Magnetism	gauss	G
Electric charge	franklin	Fr

Table 2-2 lists the primary units of measurement in the MKS system, along with their abbreviations.

Table 2-2	Units of Measurement in the MKS System	
Measurement	Unit	Abbreviation
Length	meter	m
Mass	kilogram	kg
Time	second	s
Force	Newton	N
Energy	Joule	J
Pressure	Pascal	Pa
Electric current	ampere	A
Magnetism	Tesla	T
Electric charge	Coulomb	C

Don't mix and match: Keeping physical units straight

Because each measurement system uses a different standard length, you can get several different numbers for one part of a problem, depending on the measurement you use. For example, if you're measuring the depth of the water in a swimming pool, you can use the MKS measurement system, which gives you an answer in meters; the CGS system, which yields a depth in centimeters; or the less common FPI system, in which case you determine the depth of the water in inches.

Suppose, however, that you want to know the pressure of the water at the bottom of the pool. You can simply use the measurement you find for the depth and input it into the appropriate equation for pressure (see Chapters 14 and 15). When working with equations, however, you must always keep one thing in mind: the measurement system.

Always remember to stick with the same measurement system all the way through the problem. If you start out in the MKS system, stay with it. If you don't, your answer will be a meaningless hodgepodge, because you're switching measuring sticks for multiple items as you try to arrive at a single answer. Mixing up the measurements causes problems — imagine baking a cake where the recipe calls for two cups of flour, but you use two liters instead.

Over the years, I've seen people mix up the measurement systems over and over and then scratch their heads when their answers come out wrong. Sure, they had noble intentions, and everything about their solutions was great — and sure, they had masterful insights, masterful applications, and masterful egos. But they also had the wrong answers.

Suppose the solution to a test problem is 15 kilogram-meters per second^2, but a student comes up with the result 1,500 kilogram-centimeters per second^2. The answer is wrong not because of an error in understanding, but because the answer is in the wrong measurement system.

From meters to inches and back again: Converting between units

Physicists use various measurement systems to record numbers from their observations. But what happens when you have to convert between those systems? Physics problems sometimes try to trip you up here, giving you the data you need in mixed units: centimeters for this measurement but meters for that measurement — and maybe even mixing in inches as well. Don't be fooled. You have to convert *everything* to the same measurement system before you can proceed. How do you convert in the easiest possible way? You use conversion factors. For an example, consider the following problem.

Passing another state line, you note that you've gone 4,680 miles in exactly three days. Very impressive. If you went at a constant speed, how fast were you going? As I discuss in Chapter 3, the physics notion of speed is just as you may expect — distance divided by time. So, you calculate your speed as follows:

$$\frac{4{,}680 \text{ miles}}{3 \text{ days}} = 1{,}560 \text{ miles/day}$$

Your answer, however, isn't exactly in a standard unit of measure. You want to know the result in a unit you can get your hands on — for example, miles per hour. To get miles per hour, you need to convert units.

To convert between measurements in different measuring systems, you can multiply by a conversion factor. A *conversion factor* is a ratio that, when multiplied by the item you're converting, cancels out the units you don't want and leaves those that you do. The conversion factor must equal 1.

In the preceding problem, you have a result in miles per day, which is written as miles/day. To calculate miles per hour, you need a conversion factor that knocks days out of the denominator and leaves hours in its place, so you multiply by days per hour and cancel out days:

miles/day × days/hour = miles/hour

Your conversion factor is days per hour. When you plug in all the numbers, simplify the miles-per-day fraction, and multiply by the conversion factor, your work looks like this:

4,680 miles/3 days = 1,560 miles/1 day = 1,560 miles/day × 1 day/24 hours

Note: Words like "seconds" and "meters" act like the variables *x* and *y* in that if they're present in both the numerator and denominator, they cancel each other out.

When numbers make your head spin, look at the units

Want an inside trick that teachers and instructors often use to solve physics problems? Pay attention to the units you're working with. I've had thousands of one-on-one problem-solving sessions with students in which we worked on homework problems, and I can tell you that this is a trick instructors use all the time.

As a simple example, say you're given a distance and a time, and you have to find a speed. You can cut through the wording of the problem immediately, because you know that distance (for example, meters) divided by time (for example, seconds) gives you speed (meters/second).

As the problems get more complex, however, more items will be involved — say, for example,

a mass, a distance, a time, and so on. You find yourself glancing over the words of a problem to pick out the numeric values and their units. Have to find an amount of energy? As I discuss in Chapter 10, energy is mass times distance squared over time squared, so if you can identify these items in the question, you know how they're going to fit into the solution, and you won't get lost in the numbers.

The upshot is that units are your friends. They give you an easy way to make sure you're headed toward the answer you want. So, when you feel too wrapped up in the numbers, check the units to make sure you're on the right path.

Note that because there are 24 hours in a day, the conversion factor equals exactly 1, as all conversion factors must. So, when you multiply 1,560 miles/day by this conversion factor, you're not changing anything — all you're doing is multiplying by 1.

When you cancel out days and multiply across the fractions, you get the answer you've been searching for:

$$\frac{1,560 \text{ miles}}{\text{day}} \times \frac{1 \text{ day}}{24 \text{ hours}} = \frac{65 \text{ miles}}{\text{hour}}$$

So, your average speed is 65 miles per hour, which is pretty fast considering that this problem assumes you've been driving continuously for three days.

You don't *have* to use a conversion factor; if you instinctively know that to convert from miles per day to miles per hour you need to divide by 24, so much the better. But if you're ever in doubt, use a conversion factor and write out the calculations, because taking the long road is far better than making a mistake. I've see far too many people get everything in a problem right except for this kind of simple conversion.

Converting between hours and days is pretty easy, because you know that a day consists of 24 hours. However, not all conversions are so obvious; you may not be familiar with the CGS and MKS systems, so Table 2-3 gives you a handy list of conversions for reference (refer to Tables 2-1 and 2-2 for the abbreviations).

Table 2-3	Conversions from the MKS System to the CGS System
MKS Measurement	*CGS Measurement*
1 m	100 cm
1 km	1,000 m
1 kg	1,000 g
1 N	10^5 dynes
1 J	10^7 ergs
1 Pa	10 ba
1 A	0.1 Bi
1 T	10^4 G
1 C	2.9979×10^9 Fr

Because the difference between CGS and MKS is almost always a factor of 100, converting between the two systems is easy. For example, if you know that a ball drops 5 meters, but you need the distance in centimeters, you just multiply by 100 centimeters/1 meter to get your answer:

$$5.0 \text{ meters} \times \frac{100 \text{ centimeters}}{1 \text{ meter}} = 500 \text{ centimeters}$$

However, what if you need to convert to and from the FPI system? No problem. I include all the conversions you need in the front of this book, on the Cheat Sheet. Keep it on hand when reading through this book or when tackling physics problems on your own.

Eliminating Some Zeros: Using Scientific Notation

Physicists have a way of getting their minds into the darndest places, and those places often involve really big or really small numbers. For example, say you're dealing with the distance between the sun and Pluto, which is 5,890,000,000,000 meters. You have a lot of meters on your hands, accompanied by a lot of zeroes. Physics has a way of dealing with very large and very small numbers; to help reduce clutter and make them easier to digest, it uses *scientific notation*. In scientific notation, you express zeroes as a power of ten — to get the right power of ten, you count up all the places in front of the decimal point, from right to left, up to the place just to the right of the first digit (you don't include the first digit because you leave it in front of the decimal point in the result). So you can write the distance between the sun and Pluto as follows:

$$5{,}890{,}000{,}000{,}000 \text{ meters} = 5.89 \times 10^{12} \text{ meters}$$

Scientific notation also works for very small numbers, such as the one that follows, where the power of ten is negative. You count the number of places, moving left to right, from the decimal point to just after the first nonzero digit (again leaving the result with just one digit in front of the decimal):

$$0.0000000000000000005339 \text{ meters} = 5.339 \times 10^{-19} \text{ meters}$$

If the number you're working with is greater than ten, you'll have a positive exponent in scientific notation; if it's less than one, you'll have a negative exponent. As you can see, handling super large or super small numbers with scientific notation is easier than writing them all out, which is why calculators come with this kind of functionality already built in.

Checking the Precision of Measurements

Precision is all-important when it comes to making (and analyzing) measurements in physics. You can't imply that your measurement is more precise than you know it to be by adding too many significant digits, and you have to account for the possibility of error in your measurement system by adding a ± when necessary. The following sections delve deeper into the topics of significant digits and accuracy.

Knowing which digits are significant

In a measurement, *significant digits* are those that were actually measured. So, for example, if someone tells you that a rocket traveled 10.0 meters in 7.00 seconds, the person is telling you that the measurements are known to three significant digits (the number of digits in both of the measurements).

If you want to find the rocket's speed, you can whip out a calculator and divide 10.0 by 7.00 to come up with 1.428571429 meters per second, which looks like a very precise measurement indeed. But the result is too precise — if you know your measurements to only three significant digits, you can't say you know the answer to ten significant digits. Claiming as such would be like taking a meter stick, reading down to the nearest millimeter, and then writing down an answer to the nearest ten-millionth of a millimeter.

In the case of the rocket, you have only three significant digits to work with, so the best you can say is that the rocket is traveling at 1.43 meters per second, which is 1.428571429 rounded up to two decimal places. If you include any more digits, you claim an accuracy that you don't really have and haven't measured.

When you round a number, look at the digit to the right of the place you're rounding to. If that right-hand digit is 5 or greater, you should round up. If it's 4 or less, round down. For example, you should round 1.428 to 1.43 and 1.42 down to 1.4.

What if a passerby told you, however, that the rocket traveled 10.0 meters in 7.0 seconds? One value has three significant digits, and the other has only two. The rules for determining the number of significant digits when you have two different numbers are as follows:

 ✔ **When you multiply or divide numbers,** the result has the same number of significant digits as the original number that has the fewest significant digits.

  In the case of the rocket, where you need to divide, the result should have only two significant digits — so the correct answer is 1.4 meters per second.

✔ **When you add or subtract numbers,** line up the decimal points; the last significant digit in the result corresponds to the right-most column where *all* numbers still have significant digits.

If you have to add 3.6, 14, and 6.33, you'd write the answer to the nearest whole number — the 14 has no significant digits after the decimal place, so the answer shouldn't, either. To preserve significant digits, you should round the answer up to 24. You can see what I mean by taking a look for yourself:

```
   3.6
 +14
 + 6.33
  23.93
```

By convention, zeroes used simply to fill out values down to (or up to) the decimal point aren't considered significant. For example, the number 3,600 has only two significant digits by default. If you actually measure the value to be 3,600, of course, you'd express it as 3,600.0, with a decimal point; the final decimal point indicates that you mean all the digits are significant.

Estimating accuracy

Physicists don't always rely on significant digits when recording measurements. Sometimes, you see measurements such as

$$5.36 \pm 0.05 \text{ meters}$$

The ± part (0.05 meters in the preceding example) is the physicist's estimate of the possible error in the measurement, so the physicist is saying that the actual value is between 5.36 + 0.05 (that is, 5.41) meters and 5.36 – 0.05 (that is, 5.31 meters), inclusive. (It isn't the amount your measurement differs from the "right" answer as given in books; it's an indication of how precise your apparatus can measure — in other words, how reliable your results are as a measurement.)

Fathoming the ± fad

This ± business has become so popular that you see it all over the place now, as in a real-estate ad that announces 35± acres for sale. Sometimes, you even see real-estate ads with numbers such as ±35 acres, which makes you wonder whether the agent realizes that the ad means the actual acreage is in the range of –35 to +35 acres. What if you buy the place and it turns out to be –15 acres? Do you *owe* the agent 15 acres?

Arming Yourself with Basic Algebra

Yep, physics deals with plenty of equations, and to be able to handle them, you should know how to move the items in them around. Time to travel back to basic algebra for a quick refresher.

The following equation tells you the distance, *s*, that an object travels if it starts from rest and accelerates at *a* for a time *t*:

$$s = \tfrac{1}{2}at^2$$

Now suppose the problem actually tells you the time the object was in motion and the distance it traveled and asks you to calculate the object's acceleration. By rearranging the equation, you can solve for the acceleration:

$$a = 2s \,/\, t^2$$

In this case, you multiply both sides by 2 and divide both sides by t^2 in order to isolate the acceleration, *a,* on one side of the equation.

What if you have to solve for the time, *t?* By moving the number and variables around, you get the following equation:

$$t = \sqrt{\tfrac{2s}{a}}$$

Do you need to memorize all three of these variations on the same equation? Certainly not. You just memorize one equation that relates these three items — distance, acceleration, and time — and then rearrange the equation as needed. (For a handy list of many of the equations you should keep in mind, check out the Cheat Sheet at the front of this book.)

Tackling a Little Trig

Besides some basic algebra, you need to know a little trigonometry, including sines, cosines, and tangents, for physics problems. To find these values, you start out with a simple right triangle — take a look at Figure 2-1, which displays a triangle in all its glory, complete with labels I've provided for the sake of explanation (note in particular the angle between the two shorter sides, θ).

Figure 2-1:
A labeled
triangle that
you can use
to find trig
values.

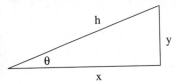

To find the trigonometric values of the triangle in Figure 2-1, you divide one side by another. You need to know the following equations, because as soon as vectors appear in Chapter 4, these equations will come in handy:

$$\sin \theta = y/r$$

$$\cos \theta = x/r$$

$$\tan \theta = y/x$$

If you're given the measure of one angle and one side of the triangle, you can find all the others. Here are some examples — they'll probably become distressingly familiar before you finish any physics course, but you *don't* need to memorize them. If you know the preceding sine, cosine, and tangent equations, you can derive the following ones as needed:

$$x = r \cos \theta = y/\tan \theta$$

$$y = r \sin \theta = x \tan \theta$$

$$r = y/\sin \theta = x/\cos \theta$$

Remember that you can go backward with the inverse sine, cosine, and tangent, which are written as $\sin^{-1}$, $\cos^{-1}$, and $\tan^{-1}$. Basically, if you input the sine of an angle into the $\sin^{-1}$ equation, you end up with the measure of the angle itself. (If you need a more in-depth refresher, check out *Trigonometry For Dummies,* by Mary Jane Sterling [Wiley].) Here are the inverses for the triangle in Figure 2-1:

$$\sin^{-1}(y/r) = \theta$$

$$\cos^{-1}(x/r) = \theta$$

$$\tan^{-1}(y/x) = \theta$$

Chapter 3

Exploring the Need for Speed

There you are in your Formula 1 racecar, speeding toward glory. You have the speed you need, and the pylons are whipping past on either side. You're confident that you can win, and coming into the final turn, you're far ahead. Or at least you think you are. Seems that another racer is also making a big effort, because you see a gleam of silver in your mirror. You get a better look and realize that you need to do something — last year's winner is gaining on you fast.

It's a good thing you know all about velocity and acceleration. With such knowledge, you know just what to do: You floor the gas pedal, accelerating out of trouble. Your knowledge of velocity lets you handle the final curve with ease. The checkered flag is a blur as you cross the finish line in record time. Not bad. You can thank your understanding of the issues in this chapter: displacement, speed, and acceleration.

You already have an intuitive feeling for what I discuss in this chapter, or you wouldn't be able to drive or even ride a bike. Displacement is all about where you are, speed is all about how fast you're going, and anyone who's ever been in a car knows about acceleration. These forces concern people every day, and physics has made an organized study of them. Knowledge of these forces has allowed people to plan roads, build spacecraft, organize traffic patterns, fly, track the motion of planets, predict the weather, and even get mad in slow-moving traffic jams.

Understanding physics is all about understanding movement, and that's the topic of this chapter. Time to move on.

Dissecting Displacement

When something moves from point A to point B, *displacement* takes place in physics terms. In plain English, displacement is a *distance.* Say, for example, that you have a fine new golf ball that's prone to rolling around by itself, shown in Figure 3-1. This particular golf ball likes to roll around on top of a large measuring stick. You place the golf ball at the 0 position on the measuring stick, as you see in Figure 3-1, diagram A.

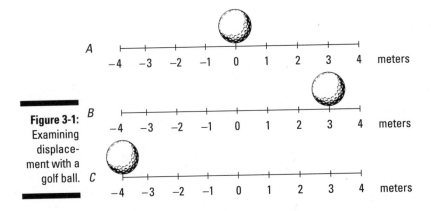

Figure 3-1: Examining displacement with a golf ball.

So far, so good. Note, however, that the golf ball rolls over to a new point, 3 meters to the right, as you see in Figure 3-1, diagram B. The golf ball has moved, so displacement has taken place. In this case, the displacement is just 3 meters to the right. Its initial position was 0 meters, and its final position is at +3 meters.

In physics terms, you often see displacement referred to as the variable s (don't ask me why). In this case, s equals 3 meters.

Like any other measurement in physics (except for certain angles), displacement always has units — usually centimeters or meters. You may also use kilometers, inches, feet, miles, or even *light years* (the distance light travels in one year, a whopper of a distance not fit for measuring with a meter stick: 5,865,696,000,000 miles, which is 9,460,800,000,000 kilometers or 9,460,800,000,000,000 meters).

Scientists, being who they are, like to go into even more detail. You often see the term s_0, which describes *initial position* (alternatively referred to as s_i; the *i* stands for *initial*). And you may see the term s_f used to describe *final position.* In these terms, moving from diagram A to diagram B in Figure 3-1, s_0 is at the 0-meter mark and s_f is at +3 meters. The displacement, s, equals the final position minus the initial position:

$$s = s_f - s_0 = +3 - 0 = 3 \text{ meters}$$

Displacements don't have to be positive; they can be zero or negative as well. Take a look at Figure 3-1, diagram C, where the restless golf ball has moved to a new location, measured as –4 meters on the measuring stick. What's the displacement here? That depends on what starting point you choose, which is always true in physics problems. The starting point is also called the *origin* (where the action originates from), and you're often free to choose the origin yourself. If you choose the 0 position on the meter stick, a common choice (because it's where the ball started), you get the following displacement:

$$s = s_f - s_0 = -4 - 0 = -4 \text{ meters}$$

Note that *s* is negative!

You can also choose origins other than the 0 position. Without seeing Figure 3-1, diagram A, for example, you may decide that the golf ball starts at the +3 meter position in Figure 3-1, diagram B, and when it moves to the –4 meter position in Figure 3-1, diagram C, you get this displacement:

$$s = s_f - s_0 = -4 - 3 = -7 \text{ meters}$$

Displacement depends on where you set your origin. Usually, your choice becomes pretty clear when you work on a problem. But what happens when the situation isn't clear and linear?

Examining axes

Motion that takes place in the world isn't always linear, like the golf ball shown in Figure 3-1. Motion can take place in two or three dimensions. And if you want to examine motion in two dimensions, you need two intersecting meter sticks, called *axes*. You have a horizontal axis — the X-axis — and a vertical axis — the Y-axis (for three-dimensional problems, watch for a third axis — the Z-axis — sticking straight up out of the paper).

Take a look at Figure 3-2, where a golf ball moves around in two dimensions. It starts at the center of the graph and moves up to the right.

In terms of the axes, the golf ball moves to +4 meters on the X-axis and +3 meters on the Y-axis, which is represented as the point (4, 3); the X measurement comes first, followed by the Y measurement: (x, y).

So what does this mean in terms of displacement? The change in the X position, Δx (Δ, the Greek letter delta, means "change in"), is equal to the final position minus the initial position. If the golf ball starts at the center of the graph — the origin of the graph, location (0, 0) — you have a change in the X location of

$$\Delta x = x_f - x_0 = +4 - 0 = 4 \text{ meters}$$

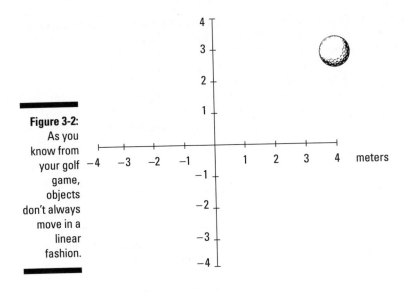

Figure 3-2:
As you
know from
your golf
game,
objects
don't always
move in a
linear
fashion.

The change in the Y location is

$$\Delta y = y_f - y_0 = +3 - 0 = 3 \text{ meters}$$

If you're more interested in figuring out the magnitude of the displacement than in the change in the X and Y location of the golf ball, that's a different story. The question now becomes: How far is the golf ball from its starting point at the center of the graph? Using the Pythagorean theorem, you can find the magnitude of the displacement of the golf ball, which is the distance it travels from start to finish. Here's how to work the equation:

$$s = \sqrt{\Delta x^2 + \Delta y^2} = \sqrt{4^2 + 3^2} = \sqrt{16 + 9} = \sqrt{25} = 5 \text{ meters}$$

So, in this case, the magnitude of the golf ball's displacement is exactly 5 meters.

The Pythagorean theorem states that the sum of the areas of the squares on the legs of a right triangle is equal to the area of the square on the hypotenuse. Visit www.cut-the-knot.org/pythagorus/index.shtml for more information.

Measuring speed

In the previous sections, you examine the motion of objects across one and two dimensions. But there's more to the story of motion than just the actual movement. When displacement takes place, it happens in a certain amount of time, which means that it happens at a certain *speed.* How long does it take the ball in Figure 3-1, for example, to move from its initial to its final position? If it takes 12 years, that makes for a long time before the figure was ready for this book. Now, 12 seconds? Sounds more like it.

Measuring how fast displacement happens is what the rest of this chapter is all about. Just as you can measure displacement, you can measure the difference in time from the beginning to the end of the motion, and you usually see it written like this:

$$\Delta t = t_f - t_o$$

Here, t_f is the final time, and t_o is the initial time, and the difference between these two is the amount of time it takes something to happen, such as a golf ball moving to its final destination. Scientists want to know all about how fast things happen, and that means measuring speed.

Speed Specifics: What Is Speed, Anyway?

You may already have the conventional idea of speed down pat, assuming you speak like a scientist:

speed = distance / time

For example, if you travel distance *s* in a time *t,* your speed, *v,* is

$$v = s / t$$

The variable *v* really stands for *velocity* (a concept I discuss in Chapter 4), but true velocity also has a direction associated with it, which speed does not. For that reason, velocity is a *vector* (you usually see the velocity vector represented as **v**). Vectors have both a magnitude and a direction, so with velocity, you know not only how fast you're going but also in what direction. Speed is only a magnitude (if you have a certain velocity vector, in fact, the speed is the magnitude of that vector), so you see it represented by the term *v* (not in bold). Displacement is also a vector, by the way, and you can read more about that in Chapter 4, too.

Phew, that was easy enough, right? Technically speaking (physicists love to speak technically), speed is the change in position divided by the change in time, so you can also represent it like this, if, say, you're moving along the X-axis:

$$v = \Delta x / \Delta t = (x_f - x_o) / (t_f - t_o)$$

You don't often see speed represented on graphs in the real world, however. Speed can take many forms, which you find out about in the following sections.

Reading the speedometer: Instantaneous speed

You already have an idea of what speed is; it's what you measure on your car's speedometer, right? When you're tooling along, all you have to do to see your speed is look down at the speedometer. There you have it: 75 miles per hour. Hmm, better slow it down a little — 65 miles per hour now. You're looking at your speed at this particular moment. In other words, you see your *instantaneous speed*.

Instantaneous speed is an important term in understanding the physics of speed, so keep it in mind. If you're going 65 mph right now, that's your instantaneous speed. If you accelerate to 75 mph, that becomes your instantaneous speed. Instantaneous speed is your speed at a particular instant of time. Two seconds from now, your instantaneous speed may be totally different.

Staying steady: Uniform speed

What if you keep driving 65 miles per hour forever? You achieve *uniform speed* in physics (also called *constant speed*), and it may be possible in the western portion of the United States, where the roads stay in straight lines for a long time and you don't have to change your speed.

Uniform motion is the simplest speed variation to describe, because it never changes.

Swerving back and forth: Nonuniform motion

Nonuniform motion varies over time; it's the kind of speed you encounter more often in the real world. When you're driving, for example, you change speed often, and your changes in speed come to life in an equation like this, where v_f is your final speed and v_o is your original speed:

$$\Delta v = v_f - v_o$$

The last part of this chapter is all about acceleration, which occurs in nonuniform motion (whether it be acceleration or deceleration).

Busting out the stopwatch: Average speed

Speed equations don't always have to be intangible ideas. You can make measuring speed more concrete. Say, for example, that you want to pound the pavement from New York to Los Angeles to visit your uncle's family, a distance of about 2,781 miles. If the trip takes you four days, what was your speed?

Speed is the distance you travel divided by the time it takes, so your speed for the trip would be

2,781 miles / 4 days = 695.3

Okay, you calculate 695.3, but 695.3 *what?*

This solution divides miles by days, so you come up with 695.3 miles per day. Not exactly a standard unit of measurement — what's that in miles per hour? To find it, you want to cancel "days" out of the equation and put in "hours" (see Chapter 2). Because a day is 24 hours, you can multiply this way (note that "days" cancel out, leaving miles over hours, or miles per hour):

(2,781 miles / 4 days) × (1 day / 24 hours) = 28.97 miles per hour

You go 28.97 miles per hour. That's a better answer, although it seems pretty slow, because when you're driving, you're used to going 65 miles per hour. You've calculated an *average speed* over the whole trip, obtained by dividing the total distance by the total trip time, which includes non-driving time.

Average speed is sometimes written as $\bar{v}$; a bar over a variable means *average* in physics terms.

Pitting average speed versus uniform motion

Average speed differs from instantaneous speed, unless you're traveling in uniform motion, in which case your speed never varies. In fact, because average speed is the total distance divided by the total time, it may be very different from your instantaneous speed.

During a trip from New York to L.A., you may stop at a hotel several nights, and while you sleep, your instantaneous speed is 0 miles per hour; yet, even at that moment, your average speed is still 28.97 miles per hour (see the previous section for this calculation)! That's because you measure average speed by dividing the whole distance, 2,781 miles, by the time the trip takes, 4 days.

Average speed also depends on the start and end points. Say, for example, that while you're driving in Ohio on your cross-country trip, you want to make a detour to visit your sister in Michigan after you drop off a hitchhiker in Indiana. Your travel path may look like the straight lines in Figure 3-3 — first 80 miles to Indiana and then 30 miles to Michigan.

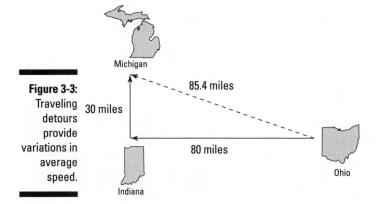

Figure 3-3:
Traveling detours provide variations in average speed.

If you drive 55 miles per hour, and you have to cover 80 + 30 = 110 miles, it takes you 2 hours. But if you calculate the speed by taking the distance between the starting point and the ending point, 85.4 miles as the crow flies, you get

85.4 miles / 2 hours = 42.7 miles per hour

You've calculated your average speed along the dotted line between the start and end points of the trip, and if that's what you really want to find, no problem. But if you're interested in your average speed along either of the two legs of the trip, you have to measure the time it takes for a leg and divide by length of that leg to get the average speed for that part of the trip.

If you move at a uniform speed, your task becomes easier. You can look at the whole distance traveled, which is 80 + 30 = 110 miles, not just 85.4 miles. And 110 miles divided by 2 hours is 55 miles per hour, which, because you travel at a constant speed, is your average speed along both legs of the trip. In fact, because your speed is constant, 55 miles per hour is also your instantaneous speed at any point on the trip.

When considering motion, it's not only speed that counts, but also direction. That's why velocity is important, because it lets you record an object's speed and its direction. Pairing speed with direction enables you to handle cases like cross-country travel, where the direction can change.

Speeding Up (or Down): Acceleration

Just like speed, you already know the basics about acceleration. *Acceleration* is how fast your speed changes. When you pass a parking lot's exit and hear squealing tires, you know what's coming next — someone is accelerating to cut you off. And sure enough, the jerk appears right in front of you, missing you by inches. After he passes, he slows down, or *decelerates,* right in front of you, forcing you to hit your brakes to decelerate yourself. Good thing you know all about physics.

Defining acceleration

In physics terms, acceleration, *a,* is the amount by which your speed changes in a given amount of time, or

$$a = \Delta v / \Delta t$$

Given the initial and final velocities, v_o and v_f, and initial and final times over which your speed changes, t_o and t_f, you can also write the equation like this:

$$a = \Delta v / \Delta t = (v_f - v_o) / (t_f - t_o)$$

Acceleration, like velocity (see Chapter 4), is actually a vector and is often written as **a,** in vector style. In other words, acceleration, like velocity and unlike speed, has a direction associated with it.

Determining the units of acceleration

You can calculate the units of acceleration easily enough by dividing speed by time to get acceleration:

$$a = (v_f - v_o) / (t_f - t_o)$$

In terms of units, the equation looks like this:

$$a = (v_f - v_o) / (t_f - t_o) = \text{distance/time} / \text{time} = \text{distance} / \text{time}^2$$

Distance over time *squared?* Don't let that throw you. You end up with time squared in the denominator because you divide velocity by time. In other words, acceleration is the rate at which your speed changes, because rates have time in the denominator.

The unit of acceleration is distance divided by time squared. So, for acceleration, you see units of meters per second^2, centimeters per second^2, miles per second^2, feet per second^2, or even kilometers per hour^2.

For example, say that you're driving at 75 miles per hour, and you see those flashing red lights in the rearview mirror. You pull over, taking about 20 seconds to come to a stop. The officer appears by your window and says, "You were going 75.0 miles per hour in a 30 mile-per-hour zone." What can you say in reply?

You can calculate how quickly you decelerated as you pulled over, which, no doubt, would impress the officer — look at you and your law-abiding tendencies! You whip out your calculator and begin entering your data. Start by converting 75.0 miles per hour into something you see more often in physics problems — centimeters per second, for example. First, you convert to miles per second:

75.0 miles per hour $\times$ (1 hour / 60 minutes) $\times$ (1 minute / 60 seconds) = 0.0208 = 2.08×10^{-2} miles per second

Next, you have to convert from miles per second to centimeters per second. You see in Chapter 2 that 1 inch = 2.54 cm, so start by converting from miles per second to inches per second:

2.08×10^{-2} mps $\times$ 5,280 feet $\times$ 12 inches = 1,318 inches

Your speed equals 1,318 inches per second, but you won't impress the cop unless you convert to centimeters per second. Grab your calculator again and do this conversion:

2.08×10^{-2} mps $\times$ 5,280 feet $\times$ 12 inches $\times$ 2.54 centimeters = 3,350 cm/s

Okay, so your original speed was 3,350 centimeters per second, and your final speed was 0 centimeters per second, and you accomplished that change in speed in 20 seconds; so what was your acceleration? You know that acceleration is the change in speed divided by the change in time, written like this:

a = Δv / Δt

Plugging in the numbers, it looks like

a = Δv / Δt = 3,350 cm/sec / 20 seconds = 168 cm/sec^2

Your acceleration was 168 cm/sec^2. But that can't be right! You may already see the problem here; take a look at the original definition of acceleration:

$$a = \Delta v / \Delta t = (v_f - v_o) / (t_f - t_o)$$

Your final speed was 0, and your original speed was 3,350 cm/sec, so plugging in the numbers here gives you this acceleration:

$$a = \Delta v / \Delta t = (v_f - v_o) / (t_f - t_o) = (0 - 3350 \text{ cm/sec}) / 20 \text{ seconds} = -168 \text{ cm/sec}^2$$

In other words, –168 cm/sec^2, *not* +168 cm/sec^2 — a big difference in terms of solving physics problems (and in terms of law enforcement). If you accelerated at +168 cm/sec^2 rather than –168 cm/sec^2, you'd end up going 150 miles per hour at the end of 20 seconds, not 0 mph. And that probably wouldn't make the cop very happy.

Now you have your acceleration. You can turn off your calculator and smile, saying, "Maybe I was going a little fast, officer, but I'm very law abiding. Why, when I heard your siren, I accelerated at –168 cm/sec^2 just in order to pull over promptly." The policeman pulls out his calculator and does some quick calculations. "Not bad," he says, impressed. And you know you're off the hook.

Like the nonvector speed, the vector acceleration takes many forms that affect your calculations in various physics situations. In different physics problems, you have to take into account whether the acceleration is positive or negative, average or instantaneous, or uniform or nonuniform. The following sections explore these variations.

Positive and negative acceleration

Don't let someone catch you on the wrong side of a numeric sign. Accelerations, like speeds, can be positive or negative, and you have to make sure you get the sign right. If you decelerate to a complete stop in a car, for example, your original speed was positive and your final speed is 0, so the acceleration is negative because a positive speed comes down to 0.

Acceleration, like speed, has a sign, as well as units.

Also, don't get fooled into thinking that a negative acceleration (deceleration) always means slowing down or that a positive acceleration always means speeding up. For example, take a look at the ball in Figure 3-4, which is happily moving in the negative direction in diagram A. In diagram B, the ball is still moving in the negative direction, but at a slower speed.

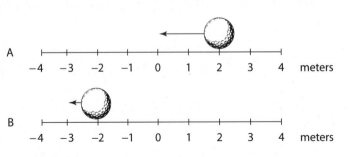

Figure 3-4:
The golf ball is traveling in the negative direction, but with a positive acceleration, so it slows down.

Because the ball's negative speed has decreased, the acceleration is *positive* during the speed decrease. In other words, to slow its negative speed, you have to add a little positive speed, which means that the acceleration is positive.

The sign of the acceleration tells you *how* the speed is changing. A positive acceleration says that the speed is increasing in the positive direction, and a negative acceleration tells you that the speed is increasing in the negative direction.

Average and instantaneous acceleration

Just as you can examine average and instantaneous speed, you can examine *average* and *instantaneous acceleration.* Average acceleration is the ratio of the change in velocity and the change in time. You calculate average acceleration, also written as $\bar{a}$, by taking the final velocity, subtracting the original velocity, and dividing the result by the total time (final time minus the original time):

$$\bar{a} = (v_f - v_o) / (t_f - t_o)$$

This equation gives you an average acceleration, but the acceleration may not have been that average value all the time. At any given point, the acceleration you measure is the instantaneous acceleration, and that number can be different from the average acceleration. For example, when you first see red flashing police lights behind you, you may jam on the brakes, which gives you a big deceleration. But as you coast to a stop, you lighten up a little, so the deceleration is smaller; however, the average acceleration is a single value, derived by dividing the overall change in speed by the overall time.

Uniform and nonuniform acceleration

Acceleration can be uniform or nonuniform. Nonuniform acceleration requires a change in acceleration. For example, when you're driving, you encounter stop signs or stop lights often, and when you decelerate to a stop and then accelerate again, you take part in nonuniform acceleration.

Other accelerations are very uniform (in other words, unchanging), such as the acceleration due to gravity on the surface of the earth. This acceleration is 9.8 meters per second^2 downward, toward the center of the earth, and it doesn't change (if it did, plenty of people would be pretty startled).

Relating Acceleration, Time, and Displacement

You deal with four quantities of motion in this chapter: acceleration, speed, time, and displacement. You work the standard equation relating displacement and time in the section on displacement to get speed:

$$v = \Delta x / \Delta t = (x_f - x_o) / (t_f - t_o)$$

And you see the standard equation relating speed and time to get acceleration from the speed section:

$$a = \Delta v / \Delta t = (v_f - v_o) / (t_f - t_o)$$

But both of these equations only go one level deep, relating speed to displacement and time and acceleration to speed and time. What if you want to relate acceleration to displacement and time?

Say, for example, you give up your oval-racing career to become a drag racer in order to analyze your acceleration down the dragway. After a test race, you know the distance you went — 0.25 miles, or about 402 meters — and you know the time it took — 5.5 seconds. So, how hard was the kick you got — the acceleration — when you blasted down the track? Good question. You want to relate acceleration, time, and displacement; speed isn't involved.

You can derive an equation relating acceleration, time, and displacement. To make this simpler, this derivation doesn't work in terms of $v_f - v_o$; you can easily subtract the initial values later.

When you're slinging around algebra, you may find it easier to work with single quantities like v rather than $v_f - v_o$, if possible. You can usually turn v into $v_f - v_o$ later, if necessary.

Not-so-distant relations

You relate acceleration, distance, and time by messing around with the equations until you get what you want. Displacement equals average speed multiplied by time:

$$s = \bar{v} \times t = \bar{v}t$$

You have a starting point. But what's the average speed during the drag race from the previous section? You started at 0 and ended up going pretty fast. Because your acceleration was constant, your speed increased in a straight line from 0 to its final value, as shown in Figure 3-5.

Figure 3-5:
Increasing
speed under
constant
acceler-
ation.

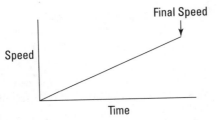

On average, your speed was half your final value, and you know this because there was constant acceleration. Your final speed was

$$v_f = a \times t = at$$

Okay, you can find your final speed, which means your average speed because it went up in a straight line, was

$$\bar{v} = \tfrac{1}{2}(a \times t)$$

So far, so good. Now you can plug this average speed into the $s = \bar{v}t$ equation and get

$$s = \bar{v}t = \tfrac{1}{2}(v_f\, t)$$

And because you know that $v_f = at$, you can get

$$s = \bar{v}t = \tfrac{1}{2}v_f t = \tfrac{1}{2}(at)(t)$$

And this becomes

$$s = \tfrac{1}{2}at^2$$

You can also put in $t_f - t_o$ rather than just plain t:

$$s = \tfrac{1}{2}\,a(t_f - t_o)^2$$

Congrats! You've worked out one of the most important equations you need to know when you work with physics problems relating acceleration, displacement, time, and speed.

Equating more speedy scenarios

What if you don't start off at zero speed, but you still want to relate acceleration, time, and displacement? What if you're going 100 miles per hour? That initial speed would certainly add to the final distance you go. Because distance equals speed multiplied by time, the equation looks like this (don't forget that this assumes the acceleration is constant):

$$s = v_0(t_f - t_o) + \tfrac{1}{2}\,(a)(t_f - t_o)^2$$

Quite a mouthful. As with other long equations, I don't recommend you memorize the extended forms of these equations unless you have a photographic memory. It's tough enough to memorize

$$s = \tfrac{1}{2}\,at^2$$

If you don't start at 0 seconds, you have to subtract the starting time to get the total time the acceleration is in effect.

If you don't start at rest, you have to add the distance that comes from the initial velocity into the result as well. If you can, it really helps to solve problems by using as much common sense as you can so you have control over everything rather than mechanically trying to apply formulas without knowing what the heck is going on, which is where errors come in.

So, what was your acceleration as you drove the drag racer I introduced in the last couple sections? Well, you know how to relate distance, acceleration, and time, and that's what you want — you always work the algebra so that you end up relating all the quantities you know to the one quantity you *don't* know. In this case, you have

$$s = \tfrac{1}{2}\,at^2$$

You can rearrange this with a little algebra; just divide both sides by t^2 and multiply by 2 to get

$$a = 2s\,/\,t^2$$

Great. Plugging in the numbers, you get

$$a = 2s / t^2 = 2(402 \text{ meters}) / (5.5 \text{ seconds})^2 = 27 \text{ meters/second}^2$$

Okay, 27 meters per second^2. What's that in more understandable terms? The acceleration due to gravity, g, is 9.8 meters per second^2, so this is about 2.7 g.

Linking Speed, Acceleration, and Displacement

Impressive, says the crafty physics textbook, you've been solving problems pretty well so far. But I think I've got you now. Imagine you're a drag racer for an example problem. I'm only going to give you the acceleration — 26.6 meters per second^2 — and your final speed — 146.3 meters per second. With this information, I want you to find the total distance traveled. Got you, huh? "Not at all," you say, supremely confident. "Just let me get my calculator."

You know the acceleration and the final speed, and you want to know the total distance it takes to get to that speed. This problem looks like a puzzler because the equations in this chapter have involved time up to this point. But if you need the time, you can always solve for it. You know the final speed, v_f, and the initial speed, v_o (which is zero), and you know the acceleration, a. Because

$$v_f - v_o = at$$

you know that

$$t = (v_f - v_o) / a = (146.3 - 0) / 26.6 = 5.5 \text{ seconds}$$

Now you have the time. You still need the distance, and you can get it this way:

$$s = \tfrac{1}{2} at^2 + v_o t$$

The second term drops out, because $v_o = 0$, so all you have to do is plug in the numbers:

$$s = \tfrac{1}{2} at^2 = \tfrac{1}{2} (26.6)(5.5)^2 = 402 \text{ meters}$$

In other words, the total distance traveled is 402 meters, or a quarter mile. Must be a quarter-mile racetrack.

If you're an equation junkie (and who isn't?), you can make this simpler on yourself with a new equation, the last one for this chapter. You want to relate distance, acceleration, and speed. Here's how it works; first, you solve for the time:

$$t = (v_f - v_o) / a$$

Because displacement $= \bar{v}t$, and $\bar{v} = \frac{1}{2}(v_o + v_f)$ when the acceleration is constant, you can get

$$s = \frac{1}{2}(v_o + v_f)t$$

Substituting for the time, you get

$$s = \frac{1}{2}(v_o + v_f)t = \frac{1}{2}(v_o + v_f)\left[(v_f - v_o) / a\right]$$

After doing the algebra, you get

$$s = \frac{1}{2}(v_o + v_f)t = \frac{1}{2}(v_o + v_f)[(v_f - v_o) / a] = (v_f^2 - v_o^2) / 2a$$

Moving the 2a to the other side of the equation, you get an important equation of motion:

$$v_f^2 - v_o^2 = 2as = 2a(x_f - x_o)$$

Whew. If you can memorize this one, you're able to relate velocity, acceleration, and distance.

You can now consider yourself a motion master.

Chapter 4

Following Directions: Which Way Are You Going?

*Y*ou have a hard time getting where you want to go — on foot, on a bike, in a car, or in a plane — if you don't know which *way* to go. Distance isn't all that matters; you need to know how far and in which direction you must go. Chapter 3 is all about forces like displacement, speed, and acceleration, with equations like $s = v_0(t_f - t_o) + \frac{1}{2} a(t_f - t_o)^2$, giving you your distance. Working with these forces, you get answers like 27 meters per second^2 or 42.7 miles per hour. These figures are fine, but what about the direction?

In the real world, you need to know which way you're going. That's what *vectors* are all about. Too many people who've had tussles with vectors decide they don't like them, which is a mistake — vectors are easy when you get a handle on them, and you're going to get a handle on them in this chapter. I break down vectors from top to bottom and relate the forces of motion to the concept of vectors.

Conquering Vectors

In Chapter 3, you work with simple numbers, or measurements called *magnitudes* in physics. If you measure a displacement of 3.0 meters, for example, that displacement has a magnitude of 3.0 meters. *Vectors* are the next step up; they add a direction to a magnitude. They mimic everyday life — when a person gives you directions, she may say something like, "The posse went 15 miles thataway" and point, giving you a magnitude (a measurement) and a direction (by pointing). When you're helping someone hang a door, the person may say, "Push hard to the left!" Another vector. When you swerve to avoid hitting someone in your car, you accelerate or decelerate in another direction. Yet another vector.

Plenty of situations in everyday life display vectors, such as driving directions, door-hanging instructions, or swerving to avoid an accident. And because physics mimics everyday life, plenty of concepts in physics are vectors too, like velocity, acceleration, force, and many more. For that reason, you should snuggle up to vectors, because you see them in just about any physics course you take. Vectors are fundamental.

Asking for directions: Vector basics

When you have a vector, you have to keep in mind two quantities: its direction and its magnitude. Technically speaking, forces that have only a quantity, like speed, are called *scalars*. If you add a direction to a scalar, you create a vector.

Visually, you see vectors drawn as arrows in physics, which is perfect because an arrow has both a clear direction and a clear magnitude (the length of the arrow). Take a look at Figure 4-1. The arrow represents a vector that starts at the foot and ends at the head.

Figure 4-1:
The arrow, a
vector, has
both a
direction
and a
magnitude.

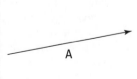

You can use vectors to represent a force, an acceleration, a velocity, and so on. In physics, you use **A** to represent a vector. In some books, you see it with an arrow on top like this:

$\vec{A}$

The arrow means that this is not only a scalar value, which would be represented by A, but also something with direction.

Say that you tell some smartypants that you know all about vectors. When he asks you to give him a vector, **A,** you give him not only its magnitude but also its direction, because you need these two bits of info together to define this vector. That impresses him to no end! For example, you may say that **A** is a vector at 15 degrees (15°) from the horizontal with a magnitude of 12 meters per second. Smartypants knows all he needs to know, including that **A** is a velocity vector.

Take a look at Figure 4-2, which features two vectors, **A** and **B.** They look pretty much the same — the same length and the same direction. In fact, these vectors are *equal.* Two vectors are equal if they have the same magnitude and direction, and you can write this as **A = B.**

Figure 4-2:
Two arrows
(and
vectors)
with the
same
magnitude
and
direction.

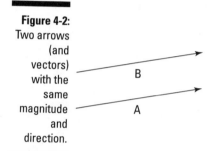

You're on your way to becoming a vector pro. Already you know that when you see the symbol **A,** you're dealing with an item that has a magnitude and a direction — or a vector — and that two vectors are equal if they have the same magnitude and direction. But there's more to come. What if, for example, someone says the hotel you're looking for is 20 miles due north *and then* 20 miles due east? How far away is the hotel, and in which direction?

Putting directions together: Adding vectors

You can add two direction vectors together, and when you do, you get a *resultant vector* — the sum of the two — that gives you the distance to your target and the direction to that target.

Assume, for example, that a passerby tells you that to get to your destination, you first have to follow vector **A** and then vector **B.** Just where is that destination? You work this problem just as you find the destination in everyday life. First, you drive to the end of vector **A,** and at that point, you drive to the end of vector **B,** just as you see in Figure 4-3.

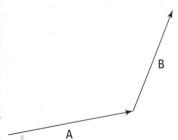

Figure 4-3:
Going from
the tail of
one vector
to the head
of a second
gets you
to your
destination.

When you get to the end of vector **B,** how far are you from your starting point? To find out, you draw a vector, **C,** from your starting point to your ending point, as you see in Figure 4-4.

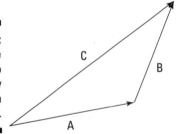

Figure 4-4:
Take the
sum of two
vectors by
creating a
new vector.

This new vector, **C,** represents the result of your complete trip, from start to finish. To get **C,** all you have to do is draw the two vectors **A** and **B** and then draw **C** as the resulting vector.

You make vector addition simple by putting one at the end of the other and drawing the new vector, or the sum, from the start of the first vector to the end of the second. In other words, **C = A + B. C** is called the sum, the result, or the resultant vector. But if having only one option bores you, there are other ways of combining vectors, too — you can subtract them if you want.

Taking distance apart: Subtracting vectors

What if someone hands you vector **C** and vector **A** from Figure 4-4 and says, "Can you get their difference?" The difference is vector **B,** because when you add vectors **A** and **B** together, you end up with vector **C.** To show the person what you mean, you explain that you subtract **A** from **C** like this: **C – A.** You don't come across vector subtraction that often in physics problems, but it does pop up, so it's worth a look.

To subtract two vectors, you put their feet (the non-pointy part of the arrow) together — you don't put the foot of one at the head of the other, as you do when adding vectors — and then draw the resultant vector, which is the difference of the two vectors, from the head of the vector you're subtracting **(A)** to the head of the vector you're subtracting it *from* **(C).** To make heads from tails, check out Figure 4-5, where you subtract **A** from **C** (in other words, **C – A**). As you can see, the result is **B**, because **C = A + B.**

Figure 4-5:
Subtracting
two vectors
by putting
their feet
together
and drawing
the result.

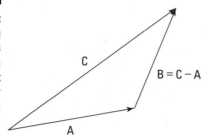

$B = C - A$

Another (and for some people easier) way to do vector subtraction is to reverse the direction of the second vector (**A** in **C – A**) and use vector addition (that is, start with the first vector [**C**] and put the reversed vector's foot [**A**] at the first vector's head, and draw the resulting vector).

As you can see, both vector addition and subtraction are possible with the same vectors in the same problems. In fact, all kinds of math operations are possible on vectors. That means that in equation form, you can play with vectors just as you can scalars, like **C = A + B, C – A = B,** and so on. This looks pretty numerical, and it is. You can get numerical with vectors just as you can with scalars.

Waxing Numerical on Vectors

Vectors may look good as arrows, but that's not exactly the most precise way of dealing with them. You can get numerical on vectors, taking them apart as you need them. Take a look at the vector addition problem, **A + B,** shown in Figure 4-6. Now that you have the vectors plotted on a graph, you can see how easy vector addition really is.

Assume that the measurements in Figure 4-6 are in meters. That means vector **A** is 1 meter up and 5 to the right, and vector **B** is 1 meter to the right and 4 up. To add them for the result, vector **C,** you add the horizontal parts together and the vertical parts together.

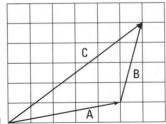

The resulting vector, **C,** ends up being 6 meters to the right and 5 meters up. You can see what that looks like in Figure 4-6 — to get the vertical part of the sum, **C,** you just add the vertical part of **A** to the vertical part of **B.** And to get the horizontal part of the sum, you add the horizontal part of **A** to the horizontal part of **B.**

If vector addition still seems cloudy, you can use a notation that was invented for vectors to help physicists and *For Dummies* readers keep it straight. Because **A** is 5 meters to the right (the positive X-axis direction) and 1 up (the positive Y-axis direction), you can express it with (x, y) coordinates like this:

A = (5, 1)

And because **B** is 1 meter to the right and 4 up, you can express it with (x, y) coordinates like this:

B = (1, 4)

Having a notation is great, because it makes vector addition totally simple. To add two vectors together, you just add their X and Y parts, respectively, to get the X and Y parts of the result:

A (5, 1) + **B** (1, 4) = **C** (6, 5)

So, the whole secret of vector addition is breaking each vector up into its X and Y parts and then adding those separately to get the resultant vector's X and Y parts? Nothing to it. Now you can get as numerical as you like, because you're just adding or subtracting numbers. It can take a little work to get those X and Y parts, but it's a necessary step. And when you have those parts, you're home free.

For another example, assume you're looking for a hotel that's 20 miles due north and then 20 miles due east. What's the vector that points at the hotel from your starting location? Taking your coordinate info into account, this is an easy problem. Say that the east direction is along the positive X-axis and that north is along the positive Y-axis. Step 1 of your travel directions is 20 miles due north, and Step 2 is 20 miles due east. You can write the problem in vector notation like this (east [positive X], north [positive Y]):

Step 1: (0, 20)

Step 2: (20, 0)

To add these two vectors together, add the coordinates:

(0, 20) + (20, 0) = (20, 20)

The resultant vector is (20, 20). It points from your starting point directly to the hotel.

For another quick numerical method, you can perform simple vector multiplication. For example, say you're driving along at 150 miles per hour eastward on a racetrack and you see a competitor in your rearview. No problem, you think; you'll just double your speed:

2(0, 150) = (0, 300)

Now you're flying along at 300 miles per hour in the same direction. In this problem, you multiply a vector by a scalar.

Breaking Up Vectors into Components

Physics problems have a way of not telling you what you want to know directly. Take a look at the first vector you see in this chapter — vector **A** in Figure 4-1. Instead of telling you that vector A is coordinate (4, 1) or something similar, a problem may say that a ball is rolling on a table at 15° with a speed of 7.0 meters per second and ask you how long it will take the ball to roll off the table's edge if that edge is 1.0 meter away to the right. Given certain information, you can find the components that make up vector problems.

Finding vector components given magnitudes and angles

You can take action to find tough vector information by breaking a vector up into its parts. When you break a vector into its parts, those parts are called its *components*. For example, in the vector (4, 1), the X-axis component is 4 and the Y-axis component is 1.

Typically, a physics problem gives you an angle and a magnitude to define a vector; you have to find the components yourself. If you know that a ball is rolling on a table at 15° with a speed of 7.0 meters per second, and you want to find out how long it will take the ball to roll off the edge 1.0 meter to the right, what you need is the X-axis direction. So, the problem breaks down to finding out how long the ball will take to roll 1.0 meter in the X direction. In order to find out, you need to know how fast the ball is moving in the X direction.

You already know that the ball is rolling at a speed of 7.0 meters per second at 15° to the horizontal (along the positive X-axis), which is a vector: 7.0 meters per second at 15° gives you both a magnitude and a direction. What you have here is a *velocity* — the vector version of speed. The ball's speed is the magnitude of its velocity vector, and when you add a direction to that speed, you get the velocity vector **v**.

Here you need not only the speed, but also the *X component* of the ball's velocity to find out how fast the ball is traveling toward the table edge. The X component is a scalar (a number, not a vector), and it's written like this: v_x. The Y component of the ball's velocity vector is v_y. So, you can say that

$$\mathbf{v} = (v_x, v_y)$$

That's how you express breaking a vector up into its components. So, what's v_x here? And for that matter, what's v_y, the Y component of the velocity? The vector has a length (7.0 meters per second) and a direction ($\theta = 15°$ to the horizontal). And you know that the edge of the table is 1.0 meter to the right. As you can see in Figure 4-7, you have to use some trigonometry (oh no!) to resolve this vector into its components. No sweat; the trig is easy after you get the angles you see in Figure 4-7 down. The magnitude of a vector **v** is expressed as v (if your physics class uses v, you use |v|), and from Figure 4-7, you can see that

$$v_x = v \cos \theta$$
$$v_y = v \sin \theta$$

Figure 4-7:
Breaking a
vector into
components
allows you
to add or
subtract
them easily.

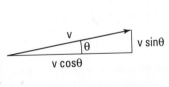

The two vector component equations are worth knowing, because you see them *a lot* in any beginning physics course. Make sure you know how they work, and always have them at your fingertips.

You can go further by relating each side of the triangle to each other side (and if you know that tan θ = sin θ / cos θ, you can derive all these from the previous two equations as required; no need to memorize all these):

$$v_x = v \cos \theta = v_y / \tan \theta$$

$$v_y = v \sin \theta = v_x \tan \theta$$

$$v = v_y / \sin \theta = v_x / \cos \theta$$

You know that $v_x = v \cos \theta$, so you can find the X component of the ball's velocity, v_x, this way:

$$v_x = v \cos \theta = v \cos 15°$$

Plugging in the numbers gives you

$$v_x = v \cos \theta = v \cos 15° = (7.0)(0.96) = 6.7 \text{ meters per second}$$

You now know that the ball is traveling at 6.7 meters per second to the right. And because the table's edge is 1.0 meter away,

$$1.0 \text{ meters} / 6.7 \text{ meters per second} = 0.15 \text{ seconds}$$

Because you know how fast the ball is going in the X direction, you now know the answer to the problem: It will take the ball 0.15 seconds to fall off the edge of the table. What about the Y component of the velocity? That's easy to find, too:

$$v_y = v \sin \theta = v \sin 15° = (7.0)(.26) = 1.8 \text{ meters per second}$$

Finding magnitudes and angles given vector components

Sometimes, you have to find the angles of a vector rather than the components. For example, assume you're looking for a hotel that's 20 miles due north and then 20 miles due east. What's the angle the hotel is at from your present location, and how far away is it? You can write this problem in vector notation, like so (see the section "Waxing Numerical on Vectors"):

Step 1: (0, 20)

Step 2: (20, 0)

When adding these vectors together, you get this result:

(0, 20) + (20, 0) = (20, 20)

The resultant vector is (20, 20). That's one way of specifying a vector — use its components. But this problem isn't asking for the results in terms of components. The question wants to know the angle the hotel is at from your present location and how far away it is. In other words, looking at Figure 4-8, the problem asks, "What's h, and what's θ?"

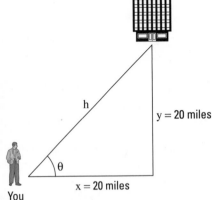

Figure 4-8:
Using the angle created by a vector to get to a hotel.

Finding h isn't so hard, because you can use the Pythagorean theorem:

$$h = \sqrt{x^2 + y^2}$$

Plugging in the numbers gives you

$$h = \sqrt{x^2 + y^2} = \sqrt{20^2 + 20^2} = 28.3 \text{ miles}$$

The hotel is 28.3 miles away. What about the angle θ? Because of your superior knowledge of trigonometry, you know that

$$x = h \cos \theta$$
$$y = h \sin \theta$$

In other words, you know that

$$x / h = \cos \theta$$
$$y / h = \sin \theta$$

Now all you have to do is take the inverse sine or cosine (you don't really need to memorize these; if you just know that $x = h \cos \theta$ and $y = h \sin \theta$, you can derive these as needed):

$$\theta = \sin^{-1}(y / h)$$

$$\theta = \cos^{-1}(x / h)$$

How do you get the inverse sine ($\sin^{-1}$) or inverse cosine ($\cos^{-1}$)? Take a look at your calculator and look for the $\sin^{-1}$ and $\cos^{-1}$ buttons! Just enter the number you want the inverse sine or inverse cosine of and press the proper button. In this case, you can find the angle θ like this:

$$\theta = \sin^{-1}(y / h) = \sin^{-1}(20 / 28.3) = 45°$$

You now know all there is to know: The hotel is 28.3 miles away, at an angle of 45°. Another physics triumph!

If you're a trig master, you can use the tangent to find the angle θ; no need to take the intermediate step of finding h first:

$$y = x \tan \theta$$

$$\theta = \tan^{-1}(y / x)$$

Unmasking the Identities of Vectors

You have two ways to describe vectors to solve physics problems. The first is in terms of their X and Y components. The second is in terms of a magnitude and an angle (the angle is usually given in degrees, 0° to 360°, where 0° is along the positive X-axis). Knowing how to convert between these two forms is pretty important, because to perform vector addition, you need the component form, but physics problems usually give vectors in the magnitude/angle form.

Here's how you can convert from the magnitude/angle form of a vector to the component form:

$$h = (x, y) = (h \cos \theta, h \sin \theta)$$

This equation assumes that θ is the angle between the horizontal component and the hypotenuse (the longest side of a triangle located opposite the right angle), which you see in Figure 4-8. If you're not given that angle, you can derive it by remembering that adding the angles together in a triangle gives you a total of 180°. And in a right triangle, like the one in Figure 4-8, the right angle = 90°, so the other two angles have to add up to 90° also.

If you're given the (x, y) component form of a vector, you can get its magnitude and angle like this:

$$h = \sqrt{x^2 + y^2}$$

$$\theta = \sin^{-1}(y\,/\,h) = \cos^{-1}(x\,/\,h) = \tan^{-1}(y\,/\,x)$$

This is the kind of stuff you should make sure you have under your belt, because the conversions trip so many people up. This is where I often see people who were really into the material before start to peel away from it — all because they didn't master converting a vector into components.

You should also remember that vectors, although mainly good for solving physics problems, have identities of their own. Guess what? Just about all quantities (except speed) I cover in Chapter 3 are really vectors, not scalars. Displacement isn't a scalar — physics treats it as a vector, and as such, it has a magnitude and a direction associated with it.

Displacement is a vector

Instead of writing displacement as *s,* you should write it as **s,** a vector with both a magnitude and a direction (if you're writing on paper, you can put an arrow over the s to signify its vector status). When you're talking about displacement in the real world, direction is as important as magnitude.

For example, say your dreams have come true: You're a big-time baseball or softball hero, slugging another line drive into the outfield. You take off for first base, which you know is 90 feet away. But 90 feet in which direction? Because you know how vital physics is, you happen to know that first base is 90 feet away at a 45° angle, as you can see in Figure 4-9. Now you're set, all because you know that displacement is a vector. In this case, here's the displacement vector:

s = 90 feet at 45°

What's that in components?

s = (s cos θ, s sin θ) = (63 feet, 63 feet)

Velocity is another vector

Imagine that you just hit a ground ball on the baseball diamond and you're running along the first-base line, or the **s** vector, 90 feet at a 45° angle to the positive X-axis. But, as you run, it occurs to you to ask: "Will my velocity enable me to evade the first baseman?" A good question, because the ball is on its way from the shortstop. Whipping out your calculator, you figure that it will take you about 3 seconds to reach first base from home plate; so, what's your velocity? To find your velocity, you quickly divide the **s** vector by the time it takes to reach first base:

s / 3.0

Figure 4-9:
A baseball diamond is a series of vectors created by the X-axis and Y-axis.

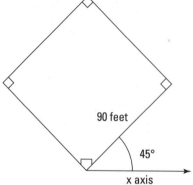

90 feet

45°

x axis

This expression represents a displacement vector divided by a time, and time is just a scalar. The result must be a vector, too. And it is: velocity, or **v:**

s / 3.0 = 90 feet at 45° / 3.0 seconds = 30 feet per second at 45° = **v**

Your velocity is 30 feet per second at 45°, and it's a vector, **v.** Dividing a vector by a scalar gives you a vector with different units and the same direction. In this case, you see that dividing a displacement vector, **s,** by a time gives you a velocity vector, **v.** It has the same magnitude as when you divided a distance by a time, but now you see a direction associated with it as well, because the displacement vector, **s,** is a vector. So, you end up with a vector result rather than the scalars you see in the Chapter 3.

With your lightning-quick physics insight, however, you decide that this velocity wouldn't be quite enough to get you to first base safely, because the first baseman already has the ball and is coming toward you. A straight direction is all wrong. What can you do? You gotta *swerve!*

Acceleration: Yep, another vector

What happens when you swerve, whether in a car or on a walk? You accelerate in a different direction. And just like displacement and velocity, acceleration is a vector, **a.**

Assume that you've just managed to hit a groundball in a softball game and you're running to first base. You figure you need the Y component of your velocity to be at least 25 feet per second to evade the first baseman, who has the ball and is waiting for you because of a poor throw down the baseline, and that you can swerve at 90° to your present path with an acceleration of 60 feet per second^2 (in a desperate attempt to dodge the first baseman). Is that acceleration going to be enough to change your velocity to what you need it to be in the tenth of a second that you have before the first baseman touches you with the ball? Sure, you're up to the challenge!

Your final time, t_f, minus your initial time, t_o, equals your change in time, Δt. You can find your change in velocity with the following equation:

$$\Delta \mathbf{v} = \mathbf{a} \Delta t$$

Now you can calculate the change in your velocity from your original velocity, as shown in Figure 4-10 (where fps = feet per second).

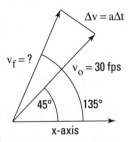

Figure 4-10:
You can use acceleration and change in time to find a change in velocity.

Finding your new velocity, $\mathbf{v_f}$, becomes an issue of vector addition. That means you have to break your original velocity, $\mathbf{v_o}$, and the change in your velocity, $\Delta \mathbf{v}$, into components. Here's what $\mathbf{v_o}$ equals:

$$\mathbf{v_o} = (v_o \cos \theta, v_o \sin \theta) = (30 \text{ fps } \cos 45°, 30 \text{ fps } \sin 45°) = (21.2 \text{ fps, } 21.2 \text{ fps})$$

You're halfway there. Now, what about $\Delta \mathbf{v}$, the change in your velocity? You know that $\Delta \mathbf{v} = \mathbf{a} \Delta t$, and that $\mathbf{a} = 60$ feet per second^2 at 90° to your present path, as shown in Figure 4-10. You can find the magnitude of $\Delta \mathbf{v}$, because $\Delta \mathbf{v} = \mathbf{a} \Delta t = (60 \text{ feet per second}^2)(0.1 \text{ seconds}) = 6$ feet per second.

But what about the angle of $\Delta \mathbf{v}$? If you look at Figure 4-10, you can see that $\Delta \mathbf{v}$ is at an angle of 90° to your present path, which is itself at an angle of 45° from the positive X-axis; therefore, $\Delta \mathbf{v}$ is at a total angle of 135° with respect to the positive X-axis. Putting that all together means that you can resolve $\Delta \mathbf{v}$ into its components:

$$\Delta \mathbf{v} = 6 \text{ fps} = (\Delta v \cos 135°, \Delta v \sin 135°) = (-4.2 \text{ fps, } 4.2 \text{ fps})$$

You now have all you need to perform the vector addition to find your final velocity:

$$\mathbf{v_o} + \Delta\mathbf{v} = \mathbf{v_f} = (21.2 \text{ fps}, 21.2 \text{ fps}) + (-4.2 \text{ fps}, 4.2 \text{ fps}) = (17.0, 25.4)$$

You've done it: $\mathbf{v_f} = (17.0, 25.4)$. The Y component of your final velocity is more than you need, which is 25.0 feet per second. Having completed your calculation, you put your calculator away and swerve as planned. And, to everyone's amazement, it works — you evade the startled first baseman and make it to first base safely without going out of the baseline (some tight swerving on your part!). The crowd roars, and you tip your hat, knowing that it's all due to your superior knowledge of physics. After the roar dies down, you take a shrewd look at second base. Can you steal it at the next pitch? It's time to calculate the vectors, so you get out your calculator again (not as pleasing to the crowd).

That's how you deal with the equations in terms of vectors for displacement, velocity, and acceleration. You can turn all the equations that use these quantities as scalars in Chapter 3 into vector equations like this:

$$\mathbf{s} = \mathbf{v_0}(t_f - t_o) + \tfrac{1}{2}\,\mathbf{a}(t_f - t_o)^2$$

Notice that total displacement is a combination of where velocity, as a vector, takes you in the given time, added to the displacement you get from constant acceleration.

Sliding Along on Gravity's Rainbow: A Velocity Exercise

Although you find out how to work with gravity in Chapter 6, you should take a look here because gravity problems present good examples of working with vectors in two dimensions. Imagine a golf ball traveling horizontally at 1.0 meter per second is about to hurtle off a 5-meter cliff, as shown in Figure 4-11. The question: How far away from the base of the cliff will it hit, and what will be its total speed? First you must find out how long the golf ball will be flying.

Figure 4-11: A tumbling golf ball shows how gravity and vectors relate.

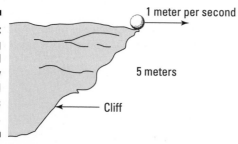

1 meter per second

5 meters

Cliff

Time to gather the facts. You know that the golf ball has the velocity vector (1, 0) and that it's flying 5.0 meters off the ground. When it falls, it comes down with a constant acceleration, g, the acceleration due to gravity, and that's 9.8 meters per second^2.

So, how can you find out how far away from the base of the cliff the golf ball will hit? One way to solve this problem is to determine how much time it will have before it hits. Because the golf ball only accelerates in the Y direction (straight down), the X component of its velocity, v_x, doesn't change, which means that the horizontal distance at which it hits will be $v_x t$, where t is the time the golf ball is in the air. Gravity is accelerating the ball as it falls, so the following equation, which relates displacement, acceleration, and time, is a good one to use:

$$\mathbf{s} = \frac{1}{2} \, \mathbf{a} t^2$$

In this case, the distance to fall is 5.0 meters, and the magnitude of a is g, the acceleration due to gravity (9.8 meters per second^2), so the equation breaks down to

$$5.0 \text{ meters} = \frac{1}{2} \, g t^2$$

which means that the time the golf ball takes to hit the ground is

$$t = \sqrt{\frac{2(5.0)}{9.8}} \text{ seconds} = 1.0 \text{ second}$$

You find out that the golf ball will be flying for 1.0 second. Okay, that's progress. Because the X component of the ball's velocity hasn't varied during that time, you can calculate how far it goes in the X direction like this:

$$x = v_x t$$

Plugging in the numbers gives you

$$x = (1.0)(1.0) = 1.0 \text{ meter}$$

The golf ball will hit 1.0 meter from the base of the cliff.

Time to figure out what the speed of the golf ball will be when it hits. You already know half the answer, because the X component of its velocity, v_x, isn't affected by gravity, so it doesn't change. Gravity is pulling on the golf ball in the Y direction, not the X, which means that the final velocity of the golf ball will look like this: (1.0, ?). So, you have to figure out the Y component of the velocity, or the *?* business in the (1.0, ?) vector. For that, you can use the following equation:

$$\mathbf{v_f} - \mathbf{v_o} = \mathbf{a}t$$

In this case, $v_o = 0$, the acceleration is g, and you want the final velocity of the golf ball in the Y direction, so the equation looks like

$$v_y = gt$$

Plugging in the numbers gives you

$$v_y = gt = (9.8)(1.0) = 9.8 \text{ meters/second}$$

But guess what? The acceleration due to gravity, g, is also a vector, $\mathbf{g}$. That makes sense, because $\mathbf{g}$ is an acceleration. This vector happens to point to the center of the earth — that is, in the negative Y direction, and on the surface of the earth, its value is -9.8 meters per second^2.

The negative sign here indicates that $\mathbf{g}$ is pointing downward, toward negative Y. So, the real result is

$$v_y = gt = (-9.8)(1.0) = -9.8 \text{ meters/second}$$

The final velocity vector of the golf ball when it hits the ground is (1.0, -9.8) meters per second. You still need to find the golf ball's speed when it hits, which is the magnitude of its velocity. You can figure that out easily enough:

$$v_f = \sqrt{1.0^2 + 9.8^2} = 9.9 \text{ meters per second}$$

You've triumphed! The golf ball will hit 1.0 meter from the base of the cliff, and its speed at that time will be 9.9 meters per second. Not bad.

Part II

May the Forces of Physics Be with You

In this part . . .

Part II gives you the lowdown on famous laws like "For every action, there is an equal and opposite reaction." The subject of forces is where Isaac Newton gets to shine. His laws and the equations in this part allow you to predict what will happen when you apply a force to an object. Mass, acceleration, friction — all the forces are here.

Chapter 5

When Push Comes to Shove: Force

his chapter is where you find Newton's famous three laws of motion. You've heard these laws before in various forms, such as "For every action, there's an equal and opposite reaction." That's not quite right; it's more like "For every *force*, there's an equal and opposite *force*," and this chapter is here to set the record straight. In this chapter, I use Newton's laws as a vehicle to focus on force and how it affects the world.

Forcing the Issue

You can't get away from forces in your everyday world; you use force to open doors, type at a keyboard, drive a car, pilot a bulldozer through a wall, climb the stairs of the Statue of Liberty (not every one, necessarily), take your wallet out of your pocket — even to breathe or talk. You unknowingly take force into account when you cross bridges, walk on ice, lift a hot dog to your mouth, unscrew a jar's cap, or flutter your eyelashes at your sweetie. Force is integrally connected to making objects move, and physics takes a big interest in understanding how it works.

Force is *fun* stuff. Like other physics topics, you may assume it's difficult, but that's before you get into it. Like your old buddies displacement, speed, and acceleration (see Chapters 3 and 4), force is a vector, meaning it has a magnitude and a direction (unlike, say, speed, which just has a magnitude).

Sir Isaac Newton was the first to put the interaction among force, mass, and acceleration into equation form in the 17th century. (He's also famous for apples dropping off trees and the consequent mathematical expression of gravity; see Chapter 6, where Newton also stars.)

Newton and the speed of light

Newton's laws have been heavily revised by the likes of Albert Einstein and his theory of relativity. The theory of relativity shows how Newton's laws don't play out the same when you're moving at speeds close to the speed of light. The idea of the theory is that the speed of light is the fastest speed possible, which means that any interaction can take place only at that speed or less; therefore, when you're getting close to that speed, you have to take into account the fact that your speed is going to affect those interactions. When you measure the length of a rocketship from nose to tail when it's moving, for example, its tail will have moved appreciably by the time you can measure with your meter stick, so the length you measure will be different than if the rocket ship were standing still next to you. As you can see in Chapter 21, Einstein's theory of relativity ends up amending Newton's view of physics quite a bit.

As with all advances in physics, Newton made observations first, modeled them mentally, and then expressed those models in mathematical terms. If you have vectors under your belt (see Chapter 4), the math is very easy.

Newton expressed his model by using three assertions, which have come to be known as Newton's laws. However, the assertions aren't really laws; the idea is that they're "laws of nature," but don't forget that physics just models the world, and as such, it's all subject to later revision.

For His First Trick, Newton's First Law of Motion

Drum roll, please. Newton's laws explain what happens with forces and motion, and his first law states: "An object continues in a state of rest, or in a state of motion at a constant speed along a straight line, unless compelled to change that state by a net force." What's the translation? If you don't apply a force to an object at rest or in motion, it will stay at rest or in that same motion along a straight line. Forever.

For example, when scoring a hockey goal, the hockey puck slides toward the open goal in a straight line because the ice it slides on is nearly frictionless. If you're lucky, the puck won't come into contact with the opposing goalie's stick, which would cause it to change its motion.

In everyday life, objects don't coast around effortlessly on ice. Most objects around you are subject to friction, so when you slide a coffee mug across your desk, it glides for a moment and then slows and comes to a stop (or spills over — don't try this one at home). That's not to say Newton's first law is invalid, just that friction provides a force to change the mug's motion to stop it.

Saying that if you don't apply a force to an object in motion, it will stay in that same motion forever sounds awfully like a perpetual-motion machine. However, you just can't get away from forces that will ultimately affect an object in motion, even if that object is in interstellar space. Even in the farthest reaches of space, the rest of the mass in the universe pulls at you, if only very slightly. And that means your motion is affected. So much for perpetual motion.

What Newton's first law really says is that the only way to get something to change its motion is to use force. In other words, force is the cause of motion. It also says that an object in motion tends to stay in motion, which introduces the idea of inertia.

Getting it going: Inertia and mass

Inertia is the natural tendency of an object to stay at rest or in constant motion along a straight line. Inertia is a quality of mass, and the mass of an object is really just a measurement of its inertia. To get an object to move — that is, to change its current state of motion — you have to apply a force to overcome its inertia.

Say, for example, you're at your summer vacation house, taking a look at the two boats at your dock: a dinghy and an oil tanker. If you apply the same force to each with your foot, the boats respond in different ways. The dinghy scoots away and glides across the water. The oil tanker moves away more slowly (what a strong leg you have!). That's because they both have different masses and, therefore, different amounts of inertia. When responding to the same force, an object with little mass — and a small amount of inertia — will accelerate faster than an object with large mass, which has a large amount of inertia.

Inertia, the tendency of mass to preserve its present state of motion, can be a problem at times. Refrigerated meat trucks, for example, have large amounts of frozen meat hanging from their ceilings, and when the drivers of the trucks begin turning corners, they create a pendulum motion they can't stop in the driver's seat. Trucks with inexperienced drivers can end up tipping over because of the inertia of the swinging frozen load in the back.

Because mass has inertia, it resists changing its motion, which is why you have to start applying forces to get velocity and acceleration. Mass ties force and acceleration together.

Measuring mass

The units of mass (and, therefore, inertia) depend on your measuring system. In the meter-kilogram-second (MKS) system or International System of Units (SI), mass is measured in kilograms (under the influence of gravity, a kilogram of mass weighs about 2.205 pounds). In the centimeter-gram-second (CGS) system, the gram is featured, and it takes a thousand grams to make a kilogram.

What's the unit of mass in the foot-pound-second system? Brace yourself: It's the *slug*. Under the influence of gravity, a slug has a weight of about 32 pounds. It's equal in mass to 14.59 kilograms.

Mass isn't the same as weight. Mass is a measure of inertia; when you put that mass into a gravitational field, you get weight. So, for example, a slug is a certain amount of mass. When you subject that slug to the gravitational pull on the surface of the earth, it has weight. And that weight is about 32 pounds.

Ladies and Gentlemen, Newton's Second Law of Motion

Newton's first law is cool, but it doesn't give you much of a handle on any math, so physicists need more. And Newton delivers with his second law: When a net force, ΣF, acts on an object of mass m, the acceleration of that mass can be calculated by using the formula $\Sigma F = ma$. Translation: Force equals mass times acceleration. The Σ you see stands for "sum," so $\Sigma F = ma$ in layman's terms is "the sum of all forces on an object, or the *net force*, equals mass times acceleration."

Newton's first law — a moving body stays in motion along a straight line unless acted on by a force — is really just a special case of Newton's second law where $\Sigma F = 0$. This means that acceleration = 0, too, which is Newton's first law. For example, take a look at the hockey puck in Figure 5-1 and imagine it's sitting there all lonely in front of a net. These two should meet.

In a totally hip move, you decide to apply your knowledge of physics to this one. You whip out your copy of this book and consult what it has to say on Newton's laws. You figure that if you apply the force of your stick to the puck for a tenth of a second, you can accelerate it in the appropriate direction. You try the experiment, and sure enough, the puck flies into the net. Score! Figure 5-1 shows how you made the goal. You applied a force to the puck, which has a certain mass, and off it went — accelerating in the direction you pushed it.

Figure 5-1: Accelerating a hockey puck

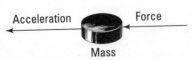

Acceleration ←   Force →

Mass

What's its acceleration? That depends on the force you apply, because $\Sigma F = ma$. But, to measure force, you have to decide on the units first.

Naming units of force

So what are the units of force? Well, $\Sigma F = ma$, so in the MKS or SI system, force must have these units:

kg-meters/sec^2

Because most people think this unit line looks a little awkward, the MKS units are given a special name: *Newtons* (named after guess who). Newtons are often abbreviated as simply N. In the CGS system, $\Sigma F = ma$ gives you the following for the units of force:

g-centimeters/sec^2

Again, it's a little awkward to rattle of these units, so CGS units are also given a special name: dynes. And 1.0×10^5 dynes = 1.0 Newton. In the foot-pound-second system, the unit of force is easy — it's just the pound. Because $\Sigma F = ma$, the pound is the same as these units:

slug-foot/sec^2

To give you a feeling for Newtons, a pound is about 4.448N.

Gathering net forces

Most books shorten $\Sigma F = ma$ to simply $F = ma$, which is what I do, too, but I must note that F stands for net force. An object you apply force to responds to the *net force* — that is, the vector sum of all the forces acting on it. Take a look, for example, at all the forces acting on the ball in Figure 5-2, represented by the arrows. Which way will the golf ball end up getting accelerated?

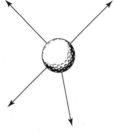

Figure 5-2:
A ball in flight faces many forces that act on it.

Because Newton's second law talks about net force, the problem becomes easier. All you have to do is add the various forces together as vectors to get the resultant, or net, force vector, ΣF, as shown in Figure 5-3. When you want to know how the ball will move, you can apply the equation $\Sigma F = ma$.

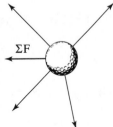

Figure 5-3:
The net force vector factors in all forces to match the ball's flight.

ΣF

Calculating displacement given a time and acceleration

Assume that you're on your traditional weekend physics data-gathering expedition. Walking around with your clipboard and white lab coat, you happen upon a football game. Very interesting, you think. In a certain situation, you observe that the football, although it starts from rest, has three players subjecting forces on it, as you see in Figure 5-4.

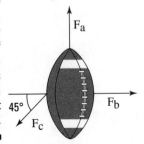

Figure 5-4:
A free body diagram of all the forces acting on a football at one time.

F_a

F_b

$45°$

F_c

In physics, Figure 5-4 is called a *free body diagram*. This kind of diagram shows all the forces acting on an object, making it easier to determine their components and find the net force.

Slipping intrepidly into the mass of moving players, risking injury in the name of science, you measure the magnitude of these forces and mark them down on your clipboard:

$F_a = 150N$

$F_b = 125N$

$F_c = 165N$

You measure the mass of the football as exactly 1.0 kilogram. Now you wonder: Where will the football be in 1 second? Here are the steps to calculate the displacement of an object in a given time with a given constant acceleration:

1. **Find the net force, ΣF, by adding all the forces acting on the object, using vector addition (see Chapter 4 for more on vector addition).**

2. **Use $\Sigma F = ma$ to determine the acceleration vector.**

3. **Use $s = v_0(t_f - t_0) + \frac{1}{2} a(t_f - t_0)^2$ to get the distance traveled in the specified time. (See Chapter 3 to find this original equation.)**

Time to get out your calculator. Because you want to relate force, mass, and acceleration together, the first order of business is to find the net force on the mass. To do that, you need to break up the force vectors you see in Figure 5-4 into their components and then add those components together to get the net force (see Chapter 4 for more info on breaking up vectors in components).

Determining F_a and F_b is easy, because F_a is straight up — along the positive Y-axis — and F_b is to the right — along the positive X-axis. That means

$$F_a = (0, 150N)$$

$$F_b = (125N, 0)$$

Finding the components of F_c is a little trickier. You need the X and Y components of this force this way:

$$F_c = (F_{cx}, F_{cy})$$

F_c is along an angle 45° with respect to the negative X-axis, as you see in Figure 5-4. If you measure all the way from the positive X-axis, you get an angle of $180° + 45° = 225°$. This is the way you break up F_c:

$$F_c = (F_{cx}, F_{cy}) = (F_c \cos \theta, F_c \sin \theta)$$

Plugging in the numbers gives you

$$F_c = (F_{cx}, F_{cy}) = (F_c \cos \theta, F_c \sin \theta) = (165N \cos 225°, 165N \sin 225°) = (-117N, -117N)$$

Look at the signs here — both components of F_c are negative. You may not follow that business about the angle of F_c being $180° + 45° = 225°$ without some extra thought, but you can always make a quick check of the signs of your vector components. F_c points downward and to the right, along the negative X- and negative Y-axes. That means that both components of this vector, F_{cx} and F_{cy}, have to be negative. I've seen many people get stuck with the wrong signs for vector components because they didn't think to make sure their numbers matched the reality.

TIP

Always compare the signs of your vector components with their actual directions along the axes. It's a quick check, and it saves you plenty of problems later.

Now you know that

$F_a = (0, 150N)$

$F_b = (125N, 0)$

$F_c = (-117N, -117N)$

You're ready for some vector addition:

$F_a = (0, 150N)$

$+ F_b = (125N, 0N)$

$+ F_c = (-117N, -117N)$

$\Sigma F = (8N, 33N)$

You calculate that the net force, ΣF, is (8N, 33N). That also gives you the direction the football will move in. The next step is to find the acceleration of the football. You know this much from Newton:

$\Sigma F = (8N, 33N) = ma$

which means that

$\Sigma F / m = (8N, 33N) / m = a$

Because the mass of the football is 1.0 kg, the problem works out like this:

$\Sigma F / m = (8N, 33N) / 1.0 \text{ kg} = (8 \text{ m/s}^2, 33 \text{ m/s}^2) = a$

You're making good progress; you now know the acceleration of the football. To find out where it will be in 1 second, you can apply the following equation (found in Chapter 3), where s is the distance:

$s = v_0(t_f - t_o) + \frac{1}{2} a(t_f - t_o)^2$

Plugging in the numbers gives you

$s = v_0(t_f - t_o) + \frac{1}{2} a(t_f - t_o)^2 = \frac{1}{2} (8 \text{ m/s}^2, 33 \text{ m/s}^2)(1.0 \text{ sec})^2 = (8 \text{ m}, 33 \text{ m})$

Well, well, well. At the end of 1 second, the football will be 8 meters along the positive X-axis and 33 meters along the positive Y-axis. You get your stopwatch out of your lab-coat pocket and measure off 1 second. Sure enough, you're right. The football moves 8 meters toward the sideline and 33 meters toward the goal line. Satisfied, you put your stopwatch back into your pocket and put a checkmark on the clipboard. Another successful physics experiment.

Calculating net force given a time and velocity

In the previous section, you calculate the displacement of an object in a given time with a given constant acceleration. But what if you want to go the other way by finding out how much force is necessary in a specific time to produce a particular velocity? Say, for example, that you want to accelerate your car from 0 to 60 miles per hour in 10 seconds; how much force is necessary? You start by converting 60 miles per hour to feet per second. First, you convert to miles per second:

(60 miles per hour) $\times$ (1 hour / 60 minutes) $\times$ (1 minute / 60 seconds) = 1.67×10^{-2} miles per second

Notice that the hours and minutes cancel out to leave you with miles and seconds for the units. Next, you take the result to feet per second:

1.67×10^{-2} miles per second $\times$ 5,280 feet per mile = 88 feet per second

You want to get to 88 feet per second in 10 seconds. If the car weighs 2,000 pounds, how much force do you need? First you find the acceleration with the following equation from Chapter 3:

$a = \Delta v / \Delta t = (v_f - v_o) / (t_f - t_o)$

Plugging in some numbers, you get

$a = \Delta v / \Delta t = (v_f - v_o) / (t_f - t_o) = $ 88 feet per second / 10 seconds = 8.8 feet per second^2

You calculate that 8.8 feet/second^2 is the acceleration you need. From Newton's second law, you know that

$\Sigma F = ma$

And you know that the weight of the car is 2,000 pounds. What's that weight in the foot-pound-second system of units, or slugs? In this system of units, you can find an object's mass given its weight by dividing by the acceleration due to gravity — 32.17 feet/second^2 (converted from 9.8 meters/second^2 — the number given to you in most physics problems):

2,000 pounds / 32 feet per second^2 = 62.5 slugs

You have all you need to know. You have to accelerate 62.5 slugs of mass by 8.8 feet per second^2; so, what force do you need? Just multiply to get your answer:

$\Sigma F = ma = $ (62.5 slugs)(8.8 feet/second^2) = 550 pounds

The car needs to supply a force of 550 pounds for those 10 seconds to accelerate you to the speed you want: 60 miles per hour.

Note that this solution ignores pesky little issues like friction and upward grade on the road; you get to those issues in Chapter 6. Even on a flat surface, friction would be large in this example, so you'd need maybe 30 percent more than that 550 pounds of force in real life.

Newton's Grand Finale: The Third Law of Motion

Newton's third law of motion is famous, especially in wrestling and drivers' education circles, but you may not recognize it in all its physics glory: Whenever one body exerts a force on a second body, the second body exerts an oppositely directed force of equal magnitude on the first body.

The more popular version of this, which I'm sure you've heard many times, is "For every action, there's an equal and opposite reaction." But for physics, it's better to express the originally intended version, and in terms of forces, not "actions" (which, from what I've seen, can apparently mean everything from voting trends to temperature forecasts).

For example, say that you're in your car, speeding up with constant acceleration. To do this, your car has to exert a force against the road; otherwise, you wouldn't be accelerating. And the road has to exert the same force on your car. You can see what this tug-of-war looks like, tire-wise, in Figure 5-5.

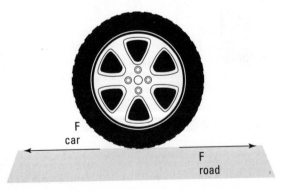

Figure 5-5:
The net forces acting on a car tire during acceleration.

If your car doesn't exert the same force against the road as the road exerts against your car, you would slip, as when the road is covered by ice and you can't get enough friction to make it exert as much force against your tires as they exert against the road. In this case, the two forces balance out, but that doesn't mean no movement takes place. Because a force acts on your car, it accelerates. That's where the force your car exerts goes — it causes your car to accelerate.

So why doesn't the road move? After all, for every force on a body, there's an equal and opposite force, so the road feels some force, too. You accelerate . . . shouldn't the road accelerate in the opposite direction? Believe it or not, it does; Newton's law is in full effect. Your car pushes the earth, affecting the motion of the earth in just the tiniest amount. Given the fact that the earth is about 6,000,000,000,000,000,000,000 times as massive as your car, however, any effects aren't too noticeable.

Tension shouldn't cause stiff necks: Friction in Newton's third law

When a hockey player slaps a puck, the puck accelerates away from the spot of contact, and so does the hockey player. If hockey pucks weighed 1,000 pounds, you'd notice this effect much more; in fact, the puck wouldn't move much at all, but the player would hurtle off in the opposite direction after striking it. (More on what happens in this case in Part III of this book.)

For fantasy physics purposes, say that the hockey game ends, and you get the job of dragging that 1,000-pound hockey puck off the rink. You use a rope to do the trick, as shown in Figure 5-6.

Figure 5-6: Pulling a 1,000-pound puck with a rope to exert equal force on both ends.

1,000-pound hockey puck

Force friction

Force rope

Ice rink

Physics problems are very fond of using ropes, including ropes with pulleys, because with ropes, the force you apply at one end is the same as the force that the rope exerts on what you tie it to at the other end.

In this case, the 1,000-pound hockey puck will have some friction that resists you — not a terrific amount, given that it slides on top of ice, but still, some. So, the net force on the puck is

$$\Sigma F = F_{rope} - F_{friction}$$

Because F_{rope} is greater than $F_{friction}$, the puck will start to move. In fact, if you pull on the rope with a constant force, the puck will accelerate, because the net force equals its mass times its acceleration:

$$\Sigma F = F_{rope} - F_{friction} = ma$$

Because some of the force you exert on the puck goes into accelerating it and some goes into overcoming the force of friction, the force you exert on the puck is the same as the force it exerts on you (but in the opposite direction), as Newton's third law predicts:

$$F_{rope} = F_{friction} + ma$$

Take a look at a completely different situation. When you pull a rope in a pulley system to lift an object, you lift the mass if you exert enough force to overcome its weight, Mg, where g is the acceleration due to gravity at the surface of the earth, 9.8 meters per second^2 (see more discussion on this in Chapter 6). Take a look at Figure 5-7, where a rope goes over a pulley and down to a mass M.

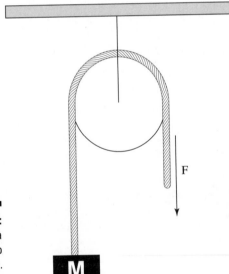

Figure 5-7:
Using a
pulley to
exert force.

The rope functions not only to transmit the force, F, you exert on the mass, M, but also to change the direction of that force, as you see in the figure. The force you exert downward is exerted on the mass upward, because the rope, going over the pulley, changes the force's direction. In this case, if F is greater than Mg, you can lift the mass; in fact, if F is greater than Mg, the mass will accelerate upward — F = M(g + a) = mass (net acceleration) in this case.

But this force-changing use of a rope and pulley comes at a cost, because you can't cheat Newton's third law. Assume that you lift the mass and it hangs in the air. In this case, F must equal Mg to hold the mass stationary. The direction of your force is being changed from downward to upward. How does that happen?

To figure this out, consider the force that the pulley's support exerts on the ceiling. What's that force? Because the pulley isn't accelerating in any direction, you know that $\Sigma F = 0$ on the pulley. That means that all the forces on the pulley, when added up, give you 0.

From the pulley's point of view, two forces pull downward: the force F you pull with and the force Mg that the mass exerts on you (because nothing is moving at the moment). That's 2F downward. To balance all the forces and get 0 total, the pulley's support must exert a force of 2F upward.

No force can be exerted without an equal and opposite opposing force (even if some of that opposing force comes from making an object accelerate). In the previous example, the rope changes the direction of the force you apply, but not for free. In order to change the direction of your force from –F (that is, downward) to +F (upward on the mass), the pulley's support has to respond with a force of 2F.

Analyzing angles and force in Newton's third law

In order to take angles into account when measuring force, you need to do a little vector addition. Take a look at Figure 5-8. Here, the mass M isn't moving, and you're applying a force F to hold it stationary. Here's the question: What force is the pulley's support exerting, and in which direction, to keep the pulley where it is?

You're sitting pretty here. Because you know that the pulley isn't moving, you know that $\Sigma F = 0$ on the pulley. So, what are the forces on the pulley? You can account for the force due to the mass's weight, Mg. Putting that in terms of vector components (see Chapter 4), it looks like this (keep in mind that the Y component of F_{mass} has to be negative, because it points downward, which is along the negative Y-axis):

$$F_{mass} = (0, -Mg)$$

You also have to account for the force of the rope on the pulley, which, because you're holding the mass stationary and the rope transmits the force you're applying, must be Mg to the right — along the positive X-axis. That force looks like this:

$$F_{rope} = (Mg, 0)$$

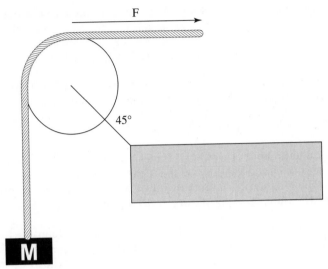

Figure 5-8:
Using a
pulley at an
angle to
keep a mass
stationary.

You can find the force exerted on the pulley by the rope and the mass by adding the vectors F_{mass} and F_{rope}:

$$F_{mass} (0, -Mg) + F_{rope} (Mg, 0) = F_{mass + rope} (Mg, -Mg)$$

$(Mg, -Mg)$ is the force exerted by the mass and the rope. You know that

$$\Sigma F = 0 = F_{mass + rope} + F_{support}$$

where $F_{support}$ is the force of the pulley's support on the pulley. This means that

$$F_{support} = -F_{mass + rope}$$

Therefore, $F_{support}$ must equal

$$-F_{mass + rope} = -(Mg, -Mg) = (-Mg, Mg)$$

As you can see by checking Figure 5-8, the directions of this vector make sense (which you should always check) — the pulley's support must exert a force to the left and upward to hold the pulley where it is.

You can also convert $F_{support}$ to magnitude and direction form (see Chapter 4), which gives you the full magnitude of the force. The magnitude is equal to

$$F_{support} = \sqrt{\left(-Mg\right)^2 + \left(Mg\right)^2} = \sqrt{2}\, Mg$$

Note that this magnitude is greater than the force you exert or the mass exerts on the pulley, because the pulley support has to change the direction of those forces.

What about the direction of $F_{support}$? You can see from Figure 5-8 that $F_{support}$ must be to the left and up, but you should see if the math bears this out. If θ is the angle of $F_{support}$ with respect to the positive X-axis, $F_{support_x}$, the X component of $F_{support}$ must be

$$F_{support_x} = F_{support}\cos\theta$$

Therefore,

$$\theta = \cos^{-1}(F_{support_x} / F_{support})$$

You know that $F_{support_x} = -Mg$ to counteract the force you exert. Because

$$F_{support} = \sqrt{2}\,Mg$$

you can figure that

$$F_{support_x} = -Mg = \frac{-F_{support}}{\sqrt{2}}$$

Now all you have to do is plug in numbers:

$$\theta = \cos^{-1}\left(\frac{F_{support_x}}{F_{support}}\right) = \cos^{-1}\left(\frac{\frac{-F_{support}}{\sqrt{2}}}{F_{support}}\right) = \cos^{-1}\left(\frac{-F_{support}}{\sqrt{2}\cdot F_{support}}\right) = \cos^{-1}\left(\frac{-1}{\sqrt{2}}\right) = -135°$$

The direction of $F_{support}$ is $-135°$ with respect to the positive X-axis — just as you expected!

If you get confused about the signs when doing this kind of work, check your answers against the directions you know the force vectors actually go in. Pictures are worth more than a thousand words, even in physics!

Finding equilibrium

In physics, an object is in equilibrium when it has zero acceleration — when the net forces acting on it are zero. The object doesn't actually have to be at rest — it can be going 1,000 miles per hour as long as the net force on it is zero and it isn't accelerating. Forces may be acting on the object, but they all add up, as vectors, to zero.

For example, take a look at Figure 5-9, where you've started your own grocery store and bought a wire rated at 15N to hang the sign with.

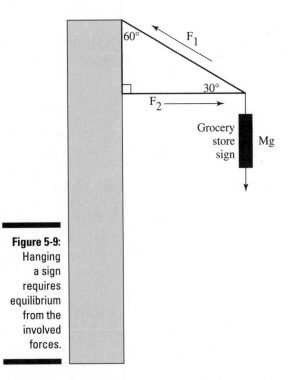

Figure 5-9:
Hanging
a sign
requires
equilibrium
from the
involved
forces.

The sign weighs only 8N, so hanging it should be no problem, right? Obviously, you can tell from my phrasing that you have a problem here. Coolly, you get out your calculator to figure out what force the wire, F_1 in the diagram, has to exert on the sign to support it. You want the sign to be at equilibrium, which means that the net force on it is zero. Therefore, the entire weight of the sign, Mg, has to be balanced out by the upward force exerted on it.

In this case, the only upward force acting on the sign is the Y component of F_1, where F_1 is the tension in the wire, as you can see in Figure 5-9. Force exerted by the horizontal brace, F2, is only horizontal, so it can't do anything for you in the vertical direction. Using your knowledge of trigonometry (see Chapter 4), you can determine from the figure that the Y component of F_1 is

$$F_{1y} = F_1 \sin 30°$$

To hold up the sign, F_{1y} must equal the weight of the sign, Mg:

$$F_{1y} = F_1 \sin 30° = Mg$$

This tells you that the tension in the wire, F_1, must be

$$F_1 = Mg / \sin 30°$$

You know that the weight of the sign is 8N, so

$$F_1 = Mg / \sin 30° = 8N / \tfrac{1}{2} = 16N$$

Uh oh. Looks like the wire will have to be able to withstand a force of 16N, not just the 15N it's rated for. The moral of this tale? You need to get a stronger wire.

Assume that you get a stronger wire. Now you may be worried about the brace that provides the horizontal force, F_2, you see diagrammed in Figure 5-9. What force does that brace have to be capable of providing? Well, you know that the figure has only two horizontal forces: F_{brace} and the X component of F_1. And you already know that $F_1 = 16N$. You have all you need to figure F_{brace}. To start, you need to determine what the X component of F_1 is. Looking at Figure 5-9 and using a little trig, you can see that

$$F_{1x} = F_1 \cos 30°$$

This is the force whose magnitude must be equal to F_{brace}:

$$F_{1x} = F_1 \cos 30° = F_{brace}$$

This tells you that

$$F_{1x} = F_1 \cos 30° = F_1(.866) = 16N(.866) = 13.9N = F_{brace}$$

The brace you use has to be able to exert a force of 13.9N.

To support a sign of just 8N, you need a wire that supports at least 16N and a brace that can provide a force of 13.9N. Look at the configuration here — the Y component of the tension in the wire has to support all the weight of the sign, and because the wire is at a pretty small angle, you need a lot of tension in the wire to get the force you need. And to be able to handle that tension, you need a pretty strong brace.

Chapter 6

What a Drag: Inclined Planes and Friction

Gravity is the main topic of this chapter. Chapter 5 shows you how much force you need to support a mass against the pull of gravity, but that's just the start. In this chapter, you find out how to handle gravity along ramps and friction in your calculations. And you also see how gravity bends the trajectory of objects flying through the air.

Don't Let It Get You Down: Dealing with Gravity

When you're on the surface of the earth, the pull of gravity is constant and equal to *mg*, where *m* is the mass of the object being pulled by gravity and *g* is the acceleration due to gravity:

$$g = 9.8 \text{ meters per second}^2 = 32.2 \text{ feet per second}^2$$

Acceleration is a vector, meaning it has a direction and a magnitude (see Chapter 4), so this equation really boils down to *g*, an acceleration straight down if you're standing on the earth. The fact that $F_{gravity} = mg$ is important, because it says that the acceleration of a falling body doesn't depend on its mass:

$$F_{gravity} = ma = mg$$

In other words,

ma = mg

Therefore, a = g, no matter the object's weight (a heavier object doesn't fall faster than a lighter one). Gravity gives any freely falling body the same acceleration downward (g near the surface of Earth).

This discussion sticks pretty close to the ground, err, Earth, where the acceleration due to gravity is constant. But Chapter 7 takes off into orbit, looking at gravity from the moon's point of view. The farther you get away from Earth, the less its gravity affects you. For the purposes of this chapter, gravity acts downward, but that doesn't mean you can only use equations like $F_{gravity} = mg$ to watch what goes up when it must come down. You can also start dealing with objects that go up at angles.

Leaning Vertical: An Inclined Plane

Plenty of gravity-oriented problems in introductory physics involve ramps, so ramps are worth taking a look at. Check out Figure 6-1. Here, a cart is about to roll down a ramp. The cart travels not only vertically but also along the ramp, which is inclined at an angle θ.

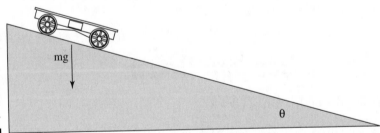

Figure 6-1:
Racing a
cart down
an inclined
ramp.

Say, for example, that θ = 30° and that the length of the ramp is 5.0 meters. How fast will the cart be going at the bottom of the ramp? Gravity will accelerate the cart down the ramp, but not the full force of gravity. Only the component of gravity acting along the ramp will accelerate the cart.

What's the component of gravity acting along the ramp if the vertical force due to gravity on the cart is F_g? Take a look at Figure 6-2, which details some of the angles and vectors (see Chapter 4 for a detailed discussion of vectors). To resolve the vector F_g along the ramp, you start by figuring out the angle between F_g and the ramp. Here's where having a knowledge of triangles comes into play (see Chapter 2) — a triangle's angles have to add up to 180°. The angle between F_g and the ground is 90°, and you know that the ramp's angle to the ground is θ. And from Figure 6-2, you know that the angle between F_g and the ramp must be 180° – 90° – θ, or 90° – θ.

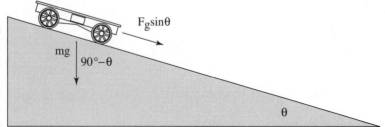

Figure 6-2:
A cart on
a ramp with
a force
vector.

Figuring out angles the easy way

Physics instructors use a top-secret technique to figure out what the angles between vectors and ramps are, and I'm here to let you in on the secret. The angles have to relate to θ in some way, so what happens if θ goes to zero? In that case, the angle between F_g from the example in the previous section and the ramp from the previous section is 90°. What happens if θ becomes 90°? In that case, the angle between F_g and the ramp is 90°.

Based on this info, you can make a pretty good case that the angle between F_g and the ramp is 90° – θ. So, when you're at a loss as how to figure out an angle with respect to another angle, let the other angle go to 90° and then 0° and see what happens. It's an easy shortcut.

Finding the component of F_g along a ramp

Now you're wondering, "What's the component of F_g along the ramp?" Now that you know that the angle between F_g and the ramp is 90° – θ (see the previous section), you can figure the component of F_g along the ramp (called *resolving* F_g along the ramp):

$$F_g \text{ along the ramp} = F_g \cos (90° – θ)$$

If you love trigonometry as much as the normal person (see Chapters 2 and 4), you may also know that (it isn't necessary to know this; the previous equation works just fine)

$$\sin θ = \cos (90° – θ)$$

Therefore,

$$F_g \text{ along the ramp} = F_g \cos (90° – θ) = F_g \sin θ \text{ along the ramp}$$

This makes sense, because when θ goes to zero, this force goes to zero as well, because the ramp is horizontal. And when θ goes to 90°, this force becomes F_g, because the ramp is vertical. The force that accelerates the cart is $F_g \sin θ$ along the ramp. What does that make the acceleration of the cart, if its mass is 800 kg? Easy enough:

$$F_g \sin θ = ma$$

Therefore,

$$a = F_g \sin θ / m$$

This becomes even easier when you remember that $F_g = mg$:

$$a = F_g \sin θ / m = mg \sin θ / m = g \sin θ$$

At this point, you know that $a = g \sin θ$ of the cart along the ramp. This equation holds for any object gravity accelerates down a ramp, if friction doesn't apply. The acceleration of an object along a ramp at angle θ to the ground is $g \sin θ$ in the absence of friction.

Figuring the speed along a ramp

All you speed freaks may be wondering: "What's the speed of the cart at the bottom of the ramp?" This looks like a job for the following equation (presented in Chapter 3):

$$v_f^2 - v_o^2 = 2as$$

The initial speed, v_o, is 0, and the distance, s, is 5.0 meters here, so you get

$$v_f^2 = 2as = 2(g \sin θ)(5.0 \text{ meters}) = 49 \text{ meter}^2 \text{ per second}^2$$

This works out to $v_f = 7.0$ meters per second, or about 15 miles per hour. Doesn't sound too fast until you try to stop an 800-kg automobile at that speed (don't try it at home). Actually, this example is a little simplified, because some of the motion goes into the angular velocity of the wheels and so on. (More on this topic in Chapter 10.)

Playing with acceleration

Quick: How fast would an ice cube on the ramp from Figures 6-1 and 6-2 go at the bottom of the ramp if friction weren't an issue? Answer: the same speed you figure in the previous section: 7.0 meters per second. The acceleration of an object along a ramp, which is at an angle θ with respect to the ground, is $g \sin θ$. The mass of the object doesn't matter — this simply takes into consideration the component of the acceleration due to gravity that acts along the ramp. And after you know the acceleration along the ramp's surface, which is a distance *s,* you can use this equation:

$$v_f^2 = 2as$$

Mass doesn't enter into it.

Getting Sticky with Friction

You know all about friction. It's the force that holds objects in motion back — or so it may seem. Actually, friction is essential for everyday living. Imagine a world without friction: no way to drive a car on the road, no way to walk on pavement, no way to pick up that ham sandwich. Friction may seem like an enemy to the hearty physics follower, but it's also your friend.

Friction comes from the interaction of surface irregularities. If you introduce two surfaces that have plenty of microscopic pits and projections, you produce friction. And the harder you press those two surfaces together, the more friction you create as the irregularities interlock more and more.

Physics has plenty to say about how friction works. For example, imagine that you decide to put all your wealth into a huge, gold ingot, which you see in Figure 6-3, only to have someone steal your fortune. The thief applies a force to the ingot to accelerate it away, as the police start after him. Thankfully, the force of friction comes to your rescue, because the thief can't accelerate away nearly as fast as he thought — all that gold drags heavily along the ground.

Figure 6-3:
The force of friction makes it tough to move large objects.

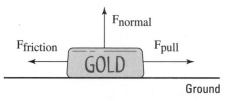

So, if you want to get quantitative here, what would you do? You'd say that the pulling force, F_{pull}, minus the force due to friction, $F_{friction}$, is equal to the net force in the X-axis direction, which gives you the acceleration in that direction:

$$F_{pull} - F_{friction} = ma$$

That looks straightforward enough. But how do you calculate $F_{friction}$? You start by calculating the normal force.

Calculating friction and the normal force

The force of friction, $F_{friction}$, always acts to oppose the force you apply when you try to move an object. Friction is proportional to the force with which an object pushes against the surface you're trying to slide it along.

As you can see in Figure 6-3, the force with which the gold ingot presses against the ground is just its weight, or *mg*. The ground presses back with the same force. The force that pushes up against the ingot is called the *normal force,* and its symbol is N. The normal force isn't necessarily the same as the force due to gravity — it's the force *perpendicular* to the surface an object is sliding on. In other words, the normal force is the force pushing the two surfaces together, and the stronger the normal force, the stronger the force due to friction.

In the case of Figure 6-3, because the ingot slides along the ground, the normal force has the same magnitude as the weight of the ingot, so F_{normal} = mg. You have the normal force, which is the force pressing the ingot and the ground together. But where do you go from there? You find the force of friction.

Conquering the coefficient of friction

The force of friction comes from the surface characteristics of the materials that come into contact. How can physics predict those characteristics theoretically? It doesn't. You see plenty of general equations that predict the general behavior of objects, like ΣF = ma (see Chapter 5). But detailed knowledge of the surfaces that come into contact isn't something that physics can come up with theoretically, so it wimps out on the theoretical part here and says that the characteristics are things you have to measure yourself.

What you measure is how the normal force (see the previous section) relates to the friction force. It turns out that to a good degree of accuracy, the two forces are proportional, and you can use a constant, μ, to relate the two:

$$F_{friction} = \mu \, F_{normal}$$

Usually, you see this equation written in the following terms:

$$F_F = \mu \, F_N$$

This equation tells you that when you have the normal force, all you have to do is multiply it by a constant to get the friction force. This constant, μ, is called the *coefficient of friction,* and it's something you measure for a particular surface — not a value you can look up in a book.

The coefficient of friction is in the range of 0.0 to 1.0. The value of 0.0 is only possible if you have a surface that has absolutely no friction at all. The maximum possible force due to friction, F_F, is F_N, the normal force. Among other things, this means that if you rely solely on the force of friction to keep you in place, you can't push a car with any force greater than your weight. That's the maximum value, and it's possible only if $\mu = 1.0$. (If you dig yourself in when pushing the car, that's a different question, because you're not just relying on the force of friction to keep you in place.)

$F_F = \mu F_N$ *isn't* a vector equation (see Chapter 4), because the force due to friction, F_F, isn't in the same direction as the normal force, F_N. As you can see in Figure 6-3, F_F and F_N are perpendicular. F_N is always at right angles to the surfaces providing the friction, because it's the force that presses the two surfaces together, and F_F is always along those surfaces, because it opposes the direction of sliding.

The force due to friction is independent of the contact area between the two surfaces, which means that even if you have an ingot that's twice as long and half as high, you still get the same frictional force when dragging it over the ground. This makes sense, because if the area of contact doubles, you may think that you should get twice as much friction. But because you've spread out the gold into a longer ingot, you halve the force on each square centimeter, because less weight is above it to push down.

Okay, are you ready to get out your lab coat and start calculating the forces due to friction? Not so fast — it turns out that you must factor in two different coefficients of friction for each type of surface.

Understanding static and kinetic friction

The two different coefficients of friction for each type of surface are a coefficient of *static friction* and a coefficient of *kinetic friction*.

The reason you have two different coefficients of friction is that you involve two different physical processes. When two surfaces are static, or not moving, and pressing together, they have the chance to interlock on the microscopic level, and that's *static friction*. When the surfaces are sliding, the microscopic irregularities don't have the same chance to connect, and you get *kinetic friction*. What this means in practice is that you must account for two different coefficients of friction for each surface: a static coefficient of friction, μ_s, and a kinetic coefficient of friction, μ_k.

Starting motion with static friction

Between static friction and kinetic friction, static friction is stronger, which means that the static coefficient of friction for a surface, μ_s, is larger than the kinetic coefficient of friction, μ_k. That makes sense, because static friction comes when the two surfaces have a chance to fully interlock on the microscopic level, and kinetic friction happens when the two surfaces are sliding, so only the more macroscopic irregularities can connect.

You create static friction when you're pushing something that starts at rest. This is the friction that you have to overcome to get something to slide.

For example, say that the static coefficient of friction between the ingot from Figure 6-3 and the ground is 0.3, and that the ingot has a mass of 1,000 kg (quite a fortune in gold). What's the force that a thief has to exert to get the ingot moving? You know from the section "Calculating friction and the normal force" that

$$F_F = \mu_s F_N$$

And because the surface is flat, the normal force — the force that drives the two surfaces together — is in the opposite direction of the ingot's weight and has the same magnitude. That means that

$$F_F = \mu_s F_N = \mu_s mg$$

where m is the mass of the ingot, and g is the acceleration due to gravity on the surface of the earth. Plugging in the numbers gives you

$$F_F = \mu_s F_N = \mu_s mg = (0.3)(1,000kg)(9.8 \text{ meters per second}^2) = 2,940N$$

The thief needs 2,940N just to get the ingot started, but what's that measurement in pounds? There are 4.448 Newtons to a pound, so that translates to about 661 pounds of force. Pretty respectable force for any thief. What happens after the burly thief gets the ingot going? How much force does he need to keep it going? He needs to look at kinetic friction.

Sustaining motion with kinetic friction

The force due to kinetic friction, which occurs when two surfaces are already sliding, isn't as strong as static friction, but that doesn't mean you can predict what the coefficient of kinetic friction is going to be, even if you know the coefficient of static friction — you have to measure both forces.

You can notice yourself that static friction is stronger than kinetic friction. Imagine that a box you're unloading onto a ramp starts to slide. To make it stop, you can put your foot in its way, and after you stop it, the box is more likely to stay put and not start sliding again. That's because static friction, which happens when the box is at rest, is greater than kinetic friction, which happens when the box is sliding.

Say that the ingot (from Figure 6-3), which weighs 1,000 kg, has a coefficient of kinetic friction, μ_k, of 0.18. How much force does the thief need to pull the ingot along during his robbery? You have all you need — the kinetic coefficient of friction:

$$F_F = \mu_k F_N = \mu_k mg$$

Putting in the numbers gives you

$$F_F = \mu_k F_N = \mu_k\, mg = (0.18)(1{,}000 \text{ kg})(9.8 \text{ meters per second}^2) = 1{,}764\text{N}$$

The thief needs 1,764N to keep your gold ingot sliding while evading the police. That converts to about 397 pounds of force (4.448N to a pound) — not exactly the kind of force you can keep going while trying to run at top speed, unless you have some friends helping you. Lucky you! Physics states that the police are able to recover your gold ingot. The cops know all about friction — taking one look at the prize, they say, "We got it back; you drag it home."

Handling uphill friction

The previous sections of this chapter deal with friction on level ground, but what if you have to drag a heavy object up a ramp? Say, for example, you have to move a refrigerator.

You want to go camping, and because you expect to catch plenty of fish, you decide to take your 100-kg refrigerator with you. The only catch is getting the refrigerator into your vehicle, as shown in Figure 6-4. The refrigerator has to go up a 30° ramp, which happens to have a static coefficient of friction of 0.2 and a kinetic coefficient of friction of 0.15 (see the previous two sections for these topics). The good news is that you have two friends to help you move the fridge. The bad news is that you can supply only 350N of force each, so your friends panic.

Figure 6-4:
You must battle different types of force and friction to push an object up a ramp.

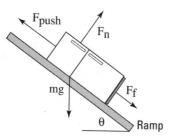

"Don't worry," you say, pulling out your calculator. "I'll check out the physics." Your two friends relax. The minimum force needed to push that refrigerator up the ramp, F_{push}, has to counter the component of the weight of the refrigerator acting along the ramp and the force due to friction. I tackle these one at a time in the following sections.

Figuring out the component weight

To start figuring the component of the weight of the refrigerator acting along the ramp, take a look at Figure 6-4. The weight of the refrigerator acts downward. The angles in a triangle formed by the ground, the ramp, and the weight vector must add up to 180°. The angle between the weight vector and the ground is 90°, and the angle between the ground and the ramp is θ, so the angle between the ramp and the weight vector is

$$180° - 90° - \theta = 90° - \theta$$

The weight acting along the ramp will be

$$mg \cos(90° - \theta) = mg \sin \theta$$

The minimum force you need to push the refrigerator up the ramp while counteracting the component of its weight along the ramp and the force of friction, F_f, is

$$F_{push} = mg \sin \theta + F_f$$

Determining the force of friction

The next question: What's the force of friction, F_f? Should you use the static coefficient of friction or the kinetic coefficient of friction? Because the static coefficient of friction is greater than the kinetic coefficient of friction, it's your best choice. After you and your friends get the refrigerator to start moving, you can keep it moving with less force. Because you're going to use the static coefficient of friction, you can get F_f this way:

$$F_F = \mu_s F_N$$

You also need the normal force, F_N, to continue (see the section "Calculating friction and the normal force" earlier in this chapter). F_N is the component of the weight perpendicular to (also called *normal* to) the ramp. You know that the angle between the weight vector and the ramp is $90° - \theta$, as shown in Figure 6-5.

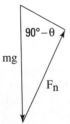

Figure 6-5:
The normal and gravitational forces acting on an object.

$90° - \theta$

mg

F_n

Using some trigonometry (see Chapter 2), you know that

$$F_N = mg \sin (90° - \theta) = mg \cos \theta$$

You can verify this by letting θ go to zero, which means that F_N becomes mg, as it should. Now you know that

$$F_{pull} = mg \sin \theta + \mu_s \, mg \cos \theta$$

All you have left is plugging in the numbers:

$$F_{pull} = mg \sin \theta + \mu_s \, mg \cos \theta = (100)(9.8)(\sin 30°) + (0.2)(100)(9.8)(\cos 30°)$$

All this breaks down to

$$(100)(9.8)(\sin 30°) + (0.2)(100)(9.8)(\cos 30°) = 490N + 170N = 660N$$

You need 660N to push the refrigerator up the ramp. In other words, your two friends, who can exert 350N each, are enough for the job. "Get started," you say, pointing confidently at the refrigerator. Unfortunately, as they chug their way up the ramp, one of them stumbles. The refrigerator begins to slide down the ramp, and they jump off, abandoning it to its fate.

Object on the loose: Calculating how far it will slide

Assuming that the ramp and the ground both have the same kinetic coefficient of friction, how far will the refrigerator your friends drop in the previous section slide? Take a look at Figure 6-6, which shows the refrigerator as it slides down the 3.0-meter ramp. As you watch with dismay, it picks up speed. A car is parked behind the ramp, only 7.5 meters away. Will the errant refrigerator smash it? You have a tense moment as you pull out your calculator to check.

Figure 6-6:
All the
forces
acting on an
object
sliding down
a ramp.

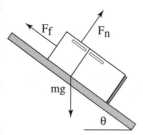

Figuring the acceleration

When an object slides downward, as you see in Figure 6-6, the forces acting on it change. With the fridge, there's no more F_{pull} force to push it up the ramp. Instead, the component of its weight acting along the ramp pulls the refrigerator downward. And while it falls down, gravity opposes that force. So, what force accelerates the refrigerator downward? You find in the section "Figuring out the component weight" that the weight acting along the ramp is

$$mg \cos(90° - \theta) = mg \sin \theta$$

And the normal force is

$$F_N = mg \sin (90° - \theta) = mg \cos \theta$$

which means that the kinetic force of friction is

$$F_F = \mu_k F_N = \mu_k mg \sin (90° - \theta) = \mu_k mg \cos \theta$$

The net force accelerating the refrigerator down the ramp, $F_{accleration}$, is the following:

$$F_{accleration} = mg \sin \theta - F_F = mg \sin \theta - \mu_k mg \cos \theta$$

Note that you subtract F_F, the force due to friction, because that force always acts to oppose the force causing the object to move. Plugging in the numbers gives you

$$F_{accleration} = (100)(9.8)(\sin 30°) - (0.15)(100)(9.8)(\cos 30°) = 490N - 127N = 363N$$

The force pulling the refrigerator down the ramp is 363N. Because the refrigerator is 100 kg, you have an acceleration of 363N / 100 kg = 3.63 meters per second^2, which acts along the entire 3.0-meter ramp. You can calculate the final speed of the refrigerator at the bottom of the ramp this way:

$$v_f^2 = 2as$$

Plugging in the numbers, you get

$$v_f^2 = 2as = 2(3.63 \text{ meters per second}^2)(3.0 \text{ meters}) = 21.8 \text{ meters}^2 \text{ per second}^2$$

This finally gives you 4.67 meters per second as the final speed of the refrigerator when it starts traveling along the street toward the parked car.

Figuring the distance traveled

With your calculations from the previous section, do you know how far the refrigerator will travel after your friends let go of it on a ramp? You start entering numbers into your calculator as your friends watch tensely.

You have a refrigerator heading down the street at 4.67 meters per second, and you need to calculate how far it's going to go. Because it's traveling along the pavement now, you need to factor in the force due to friction. And gravity will no longer accelerate the object, because the street is flat. Sooner or later, it will come to a stop. But how close will it come to a car that's parked in the street 7.5 meters away? As usual, your first calculation is the force acting on the object. In this case, you figure the force due to friction:

$$F_F = \mu_k F_N$$

Because the refrigerator is moving along a horizontal surface, the normal force, F_N, is simply the weight of the refrigerator, mg, which means the force of friction is

$$F_F = \mu_k F_N = \mu_k mg$$

Plugging in the data gives you

$$F_F = \mu_k F_N = \mu_k mg = (0.15)(100)(9.8) = 147N$$

A force of 147N acts to stop the sliding refrigerator that's now terrorizing the neighborhood. So, how far will it travel before it comes to rest? Because the force of friction in this case acts to decelerate the refrigerator, the acceleration is negative:

$$\mathbf{a} = F_F \,/\, m = -147N \,/\, 100 \text{ kg} = -1.47 \text{ meters per second}^2$$

You can find the distance through the equation

$$v_f^2 - v_o^2 = 2as$$

The distance the refrigerator slides is

$$s = (v_f^2 - v_o^2) \,/\, 2a$$

And in this case, you want the final velocity, v_f, to be zero, so this equation breaks down to

$$s = (v_f^2 - v_o^2) \,/\, 2a = (0^2 - 4.67^2) \,/\, 2(-1.47) = 7.4 \text{ meters}$$

Whew! The refrigerator slides only 7.4 meters, and the car is 7.5 meters away. With the pressure off, you watch the show as your panic-stricken friends hurtle after the refrigerator, only to see it come to a stop right before hitting the car. Just as you expected.

Determining How Gravity Affects Airborne Objects

Chapter 7 discusses gravity out in space, but I have topics to cover in this chapter when it comes to gravity near the surface of Earth. You'll come across many physics problems that deal with gravity issues, and this section is all about how what goes up must come down — the behavior of objects under the influence of constant gravitational attraction. Think of this section as a bridge between the friction and inclined plane information covered earlier in this chapter, and the gravity information to come in Chapter 7.

Going up: Maximum height

Using forces such as acceleration, gravity, and velocity, you can determine how far an object can travel straight up in the air — a good skill to have at any July 4th celebration.

Say, for example, that on your birthday, your friends give you just what you've always wanted: a cannon. It has a muzzle velocity of 860 meters per second, and it shoots cannonballs that weigh 10 kg. Anxious to show you how it works, your friends shoot it off. The only problem: The cannon is pointing straight up. Wow, you think, watching the cannonball. You wonder how high it will go, so everyone starts to guess. Because you know your physics, you can figure this one out exactly.

You know the initial velocity, v_o, of the cannonball, and you know that gravity will accelerate it downward. How can you determine how high the ball will go? At its maximum height, its speed will be zero, and then it will head down to earth again. Therefore, you can use the following equation at its highest point, where its speed will be zero:

$$v_f^2 - v_o^2 = 2as$$

where s is the cannonball's distance. This gives you

$$s = (v_f^2 - v_o^2) / 2a$$

Plugging in what you know — v_f is 0, v_o is 860 meters per second, and the acceleration is g downward (g being 9.8 meters per second^2, the acceleration due to gravity on the surface of the Earth), or $-g$ — you get this:

$$s = (v_f^2 - v_o^2) / 2a = (0^2 - 860^2) / 2(-9.8) = 3.8 \times 10^4 \text{ meters}$$

Whoa! The ball will go up 38 kilometers, or nearly 24 miles. Not bad for a birthday present. Looking around at all the guests, you wonder how long it will take the cannonball to reach its maximum height.

Floating on air: Hang time

How long would it take a cannonball shot 24 miles straight up (see the previous section) to reach its maximum height, where it's hanging at 0 velocity? You look at a somewhat similar problem in Chapter 4, where a golf ball falls off a cliff; there, you use the equation

$$s = \tfrac{1}{2}at^2$$

to determine how long the ball is in the air, given the height of the cliff. This equation is one way to come to the solution, but you have all kinds of ways to solve a problem like this. For example, you know that the speed of the cannonball at its maximum height is 0, so you can use the following equation to get the time the cannonball will take to reach its maximum height:

$$v_f = v_o + at$$

Because $v_f = 0$ and $a = -g$ (9.8 meters per second^2), it works out to this:

$$0 = v_o - gt$$

In other words,

$$t = v_o / g$$

You enter the numbers into your calculator as follows:

$$t = v_o / g = 860 / 9.8 = 88 \text{ seconds}$$

It takes 88 seconds for the cannonball to reach its maximum height. But what about the total time for the trip?

Going down: Factoring the total time

How long would it take a cannonball shot 24 miles straight into the air to complete its entire trip — up and then down, from muzzle to lawn — half of which takes 88 seconds (to reach its maximum height; see the previous sections)? Flights like the one taken by the cannonball are symmetrical; the trip up is a mirror of the trip down. The speed at any point on the way up has exactly the same magnitude as on the way down, but on the way down, the velocity is in the opposite direction. This means that the total flight time is double the time it takes the cannonball to reach its highest point, or

$$t_{total} = 2(88) = 176 \text{ seconds}$$

You have 176 seconds, or 2 minutes and 56 seconds, until the cannonball hits the ground.

Firing an object at an angle

In the previous sections, you examine a cannonball shot straight into the air with your new cannon you receive as a birthday present. Imagine now that one of your devious friends decides to fire the cannonball at an angle in order to hit target, shown in Figure 6-7. The following sections cover the cannonball's motion when you shoot at an angle.

Figure 6-7:
Shooting a
cannon at
a particular
angle with
respect to
the ground.

Breaking down a cannonball's motion into its components

How do you handle the motion of a cannonball shot at an angle? Because you can always break motion in two dimensions up into X and Y components, and because gravity acts only in the Y component, your job is easy. All you have to do is break the initial velocity into X and Y components (see Chapter 4 for the basics of this task):

$$v_x = v_o \cos \theta$$

$$v_y = v_o \sin \theta$$

These velocity components are independent, and gravity acts only in the Y direction, which means that v_x is constant; only v_y changes with time:

$$v_y = v_o \sin \theta - gt$$

If you want to know the X and Y positions of the cannonball at any time, you can easily find them. You know that X is simply

$$x = v_x t = v_o \cos \theta \, t$$

And because gravity accelerates the cannonball vertically, here's what Y looks like (the t^2 here is what gives the cannonball's trajectory in Figure 6-7 its parabolic shape):

$$y = v_y t - \tfrac{1}{2} g t^2$$

You figure out in previous sections the time it takes a cannonball to hit the ground when shot straight up:

$$t = 2v_y / g$$

Knowing that time also gives you the range of the cannon in the X direction,

$$s = v_x t = \frac{2v_x v_y}{g} = \frac{2v_o^2 \sin\theta \cos\theta}{g}$$

So there you have it — now you can figure out the range of the cannon given the speed of the cannonball and the angle at which it was shot.

Discovering the cannonball's maximum range

What's the range for your new cannon if you aim it at 45°, which will give you your maximum range? The equation you use looks like this:

$$s = v_x t = \frac{2v_x v_y}{g} = \frac{2v_o^2 \sin\theta \cos\theta}{g} = \frac{2(860)^2 \sin 45° \cos 45°}{9.8} = 75 \text{ kilometers}$$

Your range is 75 kilometers, or nearly 47 miles. Not bad.

Chapter 7

Circling around Circular Motions and Orbits

....

....

Circular motion can include rockets moving around planets, racecars whizzing around a track, or bees buzzing around a hive. The previous chapters discuss concepts like displacement, velocity, and acceleration; now you can find out how these forces work when you're moving in a circle.

You have circular equivalents for each of the forces I've mentioned, which makes handling circular motion no problem at all — you merely calculate angular displacement, angular velocity, and angular acceleration. Instead of dealing with linear displacement here, however, you deal with angular displacement as an *angle*. Angular velocity indicates what angle you sweep through in so many seconds, and angular acceleration gives you the rate of change in the angular velocity. All you have to do is take linear equations and substitute the angular equivalents: angular displacement for displacement, angular velocity for velocity, and angular acceleration for acceleration.

Time to plunge into this chapter and get dizzy with circular motion.

Staying the Course: Uniform Circular Motion

An object with *uniform circular motion* travels in a circle with a constant speed. Practical examples may be hard to come by, unless you see a racecar driver with his accelerator stuck or a clock with a seconds-hand that moves in constant motion. Take a look at Figure 7-1, where a golf ball tied to a string is whipping around in circles. The golf ball is traveling at a uniform speed as it moves around in a circle (not in a uniform velocity, because its direction changes all the time), so you can say it's traveling in uniform circular motion.

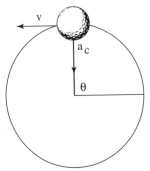

Figure 7-1: A golf ball on a string travels with constant speed, but it never makes it to the green.

Any object that travels in uniform circular motion always takes the same amount of time to move completely around the circle. That time is called its *period,* designated by T. You can easily relate the golf ball's speed to its period, because you know that the distance the golf ball must travel each time around the circle is the circumference of the circle, which, if r is the radius of the circle, is $2\pi r$. So, you can get the equation for finding an object's period by first finding its speed:

$$v = 2\pi r / T$$

If you switch T and v around, you get

$$T = 2\pi r / v$$

For example, say that you're spinning a golf ball in a circle at the end of a 1.0-meter string every half-second. How fast is the ball moving? Time to plug in the numbers:

$$v = 2\pi r / T = [2(3.14)(1.0)] / \tfrac{1}{2} = 12.6 \text{ meters/second}$$

The ball moves at a speed of 12.6 meters per second. Just make sure you have a strong string!

Changing Direction: Centripetal Acceleration

In order to keep an object moving in circular motion, its velocity constantly changes direction, as you can see in Figure 7-2. Because of this fact, acceleration is created, called *centripetal acceleration* — the acceleration needed to keep an object moving in circular motion. At any point, the velocity of the object is perpendicular to the radius of the circle.

This rule holds true for all objects: The velocity of an object in uniform circular motion is always perpendicular to the radius of the circle.

If the string holding the ball in Figure 7-2 breaks at the top, bottom, left, or right moment you see in the illustration, which way would the ball go? If the velocity points to the left, the ball would fly off to the left. If the velocity points to the right, the ball would fly off to the right. And so on. That's not intuitive for many people, but it's the kind of physics question that may come up in introductory courses.

You must also bear in mind that the velocity of an object in uniform circular motion is always at right angles to the radius of the object's path. At any one moment, the velocity points along the tiny section of the circle's circumference where the object is, so the velocity is called *tangential* to the circle.

Figure 7-2:
Velocity
constantly
changes
direction to
maintain an
object's
uniform
circular
motion.

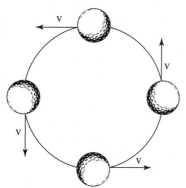

Controlling velocity with centripetal acceleration

What's special about circular motion is that when an object travels in it, its speed is constant, which means that the magnitude of the object's velocity doesn't change. Therefore, acceleration can have no component in the same direction as the velocity; if it could, the velocity's magnitude would change.

However, the velocity's direction is constantly changing — it always bends so that the object maintains movement in a constant circle. To make that happen, the object's centripetal acceleration is always concentrated toward the center of the circle, perpendicular to the object's velocity at any one time. It changes the direction of the object's velocity while keeping the magnitude of the velocity constant.

In the ball's case (see Figures 7-1 and 7-2), the string exerts a force on the ball to keep it going in a circle — a force that provides the ball's centripetal acceleration. In order to provide that force, you have to constantly pull on the ball toward the center of the circle (time for a good thought experiment: Picture what it feels like, force-wise, to whip an object around on a string). You can see the centripetal acceleration vector, a_c, in Figure 7-1.

If you accelerate the ball toward the center of the circle to provide the centripetal acceleration, why doesn't it hit your hand? The answer is that the ball is already moving at a high speed. The force, and therefore the acceleration, you provide always acts at right angles to the velocity and therefore changes only the *direction* of the velocity, not its *magnitude*.

Finding the magnitude of the centripetal acceleration

You always have to accelerate an object toward the center of the circle to keep it moving in circular motion. So, can you find the magnitude of the acceleration you create? No doubt. If an object is moving in uniform circular motion at speed v and radius r, you can find the centripetal acceleration with the following equation:

$$a_c = v^2 / r$$

For a practical example, imagine you're driving around curves at a high speed. For any constant speed, you can see from the equation $a_c = v^2 / r$ that the centripetal acceleration is inversely proportional to the radius of the curve, which you know from experience. On tighter curves, your car needs to provide a greater centripetal acceleration.

Pulling Toward the Center: Centripetal Force

When you're driving a car, you create centripetal acceleration by the friction of your tires on the road. How do you know what force you need to create to turn the car at a given speed and turning radius? That depends on the *centripetal force* — the center-seeking, inward force needed to keep an object moving in uniform circular motion.

You may have an easy time swinging a ball on a string in a circle (see Figure 7-1), but if you replace the small ball with a cannonball, watch out. You now have to whip 10 kg around on the end of a 1.0-meter string every half second. As you can tell, you need a heck of a lot more force.

Because force equals mass times acceleration, F = ma, and because centripetal acceleration is equal to v^2 / r (see the previous section), you can determine the centripetal force needed to keep an object moving in uniform circular motion with mass *m,* speed *v,* and radius *r* with the following equation:

$$F_c = mv^2 / r$$

This equation tells you how much force you need to move a given object in a circle at a given radius and speed. (***Note:*** Objects moving in circles with the same radius can have different speeds, but you need more force for faster speeds.) For example, the ball from Figures 7-1 and 7-2 is moving at 12.6 meters per second on a 1.0-meter string — how much force do you need to make a 10-kg cannonball move in the same circle at the same speed? Here's what the equation looks like:

$$F_c = mv^2 / r = [(10)(12.6)^2] / 1.0 = 1{,}590 \text{ Newtons}$$

You need about 1,590N, or about 357 pounds of force (4.448 Newtons in a pound; see Chapter 5 for more on Newtons). Pretty hefty, if you ask me; I just hope your arms can take it.

Is centripetal force actually centripetally-*needed* force?

Centripetal force isn't some new force that appears out of nowhere when an object travels in a circle; it's the force the object *needs* to keep traveling in that circle. For that reason, it may be better to refer to it as *centripetal-needed force.* I've seen countless people confused because they think centripetal force is a new force that just magically happens. The confusion is understandable, because all the other forces that you study — such as the force of gravity or friction — are real forces that apply to various objects.

Centripetal force, on the other hand, is the force you *need* to keep the object going in a circle — and to keep it going, the vector sum of all the other forces on that object must at least give you the centripetal force that you need. In other words, when you push an object with a force of 5.0N, that object feels a real force of 5.0N. But if you want to make an object travel in a circle, and it needs a force of 5.0N to perform the task, it won't travel in a circle until you apply the needed force.

Negotiating Curves and Banks: Centripetal Force through Turns

Imagine that you're driving a car, riding a bike, or jogging on a surface and you come to a banked curve. The steeper the bank of the surface, the more centripetal force you need to steer a car, a bike, or yourself around it at a high speed. The force you need comes from the friction of the tires with the surface or from the friction of your shoes against the ground; if the surface is covered with a foreign substance such as ice, you produce less friction, and you can't turn as safely at high speeds.

For example, say you're sitting in the passenger seat of the car, approaching a tight turn with a 10.0-meter radius. You know that the coefficient of static friction (see Chapter 6) is 0.8 on this road (you use the coefficient of static friction because the tires aren't slipping on the road's surface) and that the car has a mass of about 1,000 kg. What's the maximum speed the driver can go and still keep you safe? You get out your calculator as the driver turns to you with raised eyebrows. The frictional force needs to supply the centripetal force, so you come up with the following:

$$F_c = mv^2 / r = \mu_s mg$$

where m is the mass of the car, v is the velocity, r is the radius, μ is the coefficient of static friction, and g is the acceleration due to gravity, 9.8 meters per second^2. Solving for the speed on one side of the equation gives you

$$v = \sqrt{\mu_s gr}$$

This looks simple enough — you just plug in the numbers to get

$$v = \sqrt{\mu_s gr} = \sqrt{(0.8)(9.8)(10.0)} = 8.9 \text{ meters per second}$$

You calculate 8.9 meters per second, or about 20 miles per hour. You look at the speedometer and see a speed of 18 miles per hour. You can negotiate the turn safely at your present speed. However, you can't expect everyone to slow dramatically when driving on a highway, which is why highway curves are banked to make the turning easier. Some of your car's weight helps provide some of the needed centripetal force.

Take a look at Figure 7-3, which shows your car banking around a turn. What should the angle θ be if drivers go around the 200-meter-radius turn at 60 miles per hour? The engineers can make the driving experience enjoyable if they bank the turn so that drivers garner the centripetal force needed to go around the turn entirely by the *component* of their cars' weights acting toward the center of the turn's circle. That component is $F_n \sin \theta$ (F_n is the normal force; see Chapter 6), so

$$F_c = F_n \sin \theta = mv^2 / r$$

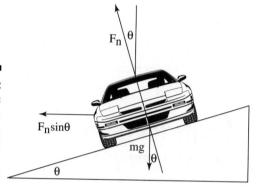

Figure 7-3:
The forces acting on a car banking around a turn.

F_n

θ

$F_n \sin\theta$

mg

θ

θ

To find the centripetal force, you need the normal force, F_n. If you look at Figure 7-3, you can see that F_n is a combination of the centripetal force due to the car banking around the turn and the car's weight. The purely vertical component of F_n must equal *mg*, because no other forces are operating vertically, so

$$F_n \cos \theta = mg$$

or

$$F_n = mg / \cos \theta$$

Plugging this result into the equation for centripetal force gives you

$$F_c = F_n \sin \theta = (mg \sin \theta / \cos \theta) = mv^2 / r$$

Because $\sin \theta$ divided by $\cos \theta = \tan \theta$, you can also write this as

$$F_c = F_n \sin \theta = (mg \sin \theta) / \cos \theta = mg \tan \theta = mv^2 / r$$

The equation finally breaks down to

$$\theta = \tan^{-1}(v^2 / gr)$$

You don't have to memorize this result, in case you're panicking. You can always come up with the equation if you're given some information. $\theta = \tan^{-1}(v^2 / gr)$ is the kind of equation used by highway engineers when they have to bank curves (notice that the mass of the car cancels out, meaning that it holds for vehicles regardless of weight). To solve the problem I present earlier, you can plug in the numbers; 60 miles per hour is about 27 meters per second, and the radius of the turn is 200 meters, so:

$$\theta = \tan^{-1}(v^2 / gr) = \tan^{-1}(27^2 / [(9.8)(200)]) = 20°$$

The designers should bank the turn at about 20° to give drivers a smooth experience. Good work; you'd make an excellent highway engineer!

Getting Angular: Displacement, Velocity, and Acceleration

If you're more of a point a to point b kind of person, you can describe circular motion in a linear fashion, but it takes a little getting used to. Take a look at the ball in Figure 7-1 — it doesn't cover distance in a linear way. You can't chart the X-axis or Y-axis coordinate of the golf ball with a straight line, but its path of motion provides one coordinate that you can graph as a straight line in uniform circular motion — the angle, θ. If you graph the angle, the total angle the ball travels increases in a straight line. When it comes to circular motion, therefore, you can think of the angle, θ, just as you think of the displacement, *s*, in linear motion (see Chapter 3 for more on displacement).

The standard unit of measurement for the linear version of circular motion is the *radian,* not the degree. A full circle is made up of 2π radians, which is also 360°, so 360° = 2π radians. If you travel in a full circle, you go 360°, or 2π radians. A half-circle is π radians, and a quarter-circle is $\pi/2$ radians.

How do you convert from degrees to radians and back again? Because 360° = 2π radians (or 2 multiplied by 3.14, the rounded version of pie), you have an easy calculation. If you have 45° and want to know how many radians that translates to, just use this conversion factor:

$$45° \frac{2\pi \text{ radians}}{360°} = \frac{\pi}{4} \text{ radians}$$

You find out that 45° = $\pi/4$ radians. If you have, say, $\pi/2$ radians and want to know how many degrees that converts to, you do this conversion:

$$\pi/2 \text{ radians } (360° / 2\pi \text{ radians}) = 90°$$

You calculate that $\pi/2$ radians = 90°.

The fact that you can think of the angle, θ, in circular motion just as you think of the displacement, s, in linear motion is great, because it means you have an angular counterpart for each of the linear equations from Chapter 3. Some such linear equations include

$v = \Delta s / \Delta t$

$a = \Delta v / \Delta t$

$s = v_0(t_f - t_o) + \frac{1}{2} a(t_f - t_o)^2$

$v_f^2 - v_o^2 = 2as$

To find the angular counterpart of each of these equations, you just make substitutions. Instead of s, which you use in linear travel, you use θ, the *angular displacement*. So, what do you use in place of the velocity, v? You use the *angular velocity*, ω, or the number of radians covered in one second:

$\omega = \Delta\theta / \Delta t$

Note that the previous equation looks close to how you define linear speed:

$v = \Delta s / \Delta t$

Say, for example, that you have a ball tied to a string. What's the angular velocity of the ball if you whirl it around on the string? It makes a complete circle, 2π radians, in $\frac{1}{2}$ seconds, so its angular velocity is

$\omega = \Delta\theta / \Delta t = 2\pi$ radians $/ \frac{1}{2}$ seconds = 4π radians/second

Can you also find the acceleration of the ball? Yes, you can, by using the *angular acceleration*, α. Linear acceleration is defined this way:

$a = \Delta v / \Delta t$

Therefore, you define angular acceleration this way:

$\alpha = \Delta\omega / \Delta t$

The units for angular acceleration are radians per second^2. If the ball speeds up from 4π radians per second to 8π radians per second in 2 seconds, for example, what would its angular acceleration be? Work it out by plugging in the numbers:

$\alpha = \Delta\omega / \Delta t = (8\pi - 4\pi) / 2 = 4\pi / 2 = 2\pi$ radians per second^2

Now you have the angular versions of linear displacement, s, velocity, v, and acceleration, a: angular displacement, θ, angular velocity, ω, and angular acceleration, α. You can make a one-for-one substitution in velocity, acceleration, and displacement equations (see Chapter 3) to get:

$$\omega = \Delta\theta / \Delta t$$

$$\alpha = \Delta\omega / \Delta t$$

$$\theta = \omega_0(t_f - t_o) + \tfrac{1}{2}\,\alpha(t_f - t_o)^2$$

$$\omega_f^2 - \omega_o^2 = 2\alpha\theta$$

If you need to work in terms of angle, not distance, you have the ammo to do so. To find out more about angular displacement, angular velocity, and angular acceleration, see the discussion on angular momentum and torque in Chapter 10.

Dropping the Apple: Newton's Law of Gravitation

You don't have to tie objects onto strings to observe travel in circular motion; larger bodies like planets move in circular motion, too. You don't see many strings extending from Earth to the moon to keep the moon circling around us; instead, *gravity* provides the necessary centripetal force.

Sir Isaac Newton came up with one of the heavyweight laws in physics for us: the law of universal gravitation. This law says that every mass exerts an attractive force on every other mass. If the two masses are m_1 and m_2, and the distance between them is r, the magnitude of the force is

$$F = (G\, m_1 m_2) / r^2$$

where G is a constant equal to 6.67×10^{-11} Nm^2/kg^2.

This equation allows you to figure the gravitational force between any two masses. What, for example, is the pull between the Sun and the Earth? The Sun has a mass of about 1.99×10^{30} kg, and the Earth has a mass of about 5.98×10^{24} kg; a distance of about 1.50×10^{11} meters separates the two bodies. Plugging the numbers into Newton's equation gives you

$$F = (G\, m_1 m_2) / r^2 = [(6.67 \times 10^{-11})(1.99 \times 10^{30})(5.98 \times 10^{24})] / (1.50 \times 10^{11})^2 = 3.52 \times 10^{22} N$$

Your answer of $3.52 \times 10^{22} N$ converts to about 8.0×10^{20} pounds of force (4.448N in a pound).

An apple a day . . .

You know the story — an apple supposedly fell on Isaac Newton's head, causing him to come up with the law of universal gravitation. But did the apple really fall? And did it actually land on his head? Although it seems likely that a falling apple set Newton off on the path of discovery, or at least brought his attention back to the issue, it apparently didn't fall on his head according to recent historical accounts. However, as far as historians know, the apple tree was in Newton's mother's garden in Woolsthorpe, near Grantham in Lincolnshire. A tree grafted from the original is still alive.

For an example on the land-based end of the spectrum, say that you're out for your daily physics observations when you notice two people on a park bench, looking at each other and smiling. As time goes on, you notice that they seem to be sitting closer and closer to each other each time you take a glance. In fact, after a while, they're sitting right next to each other. What could be causing this attraction? If the two lovebirds weigh about 75 kg each, what's the force of gravity pulling them together, assuming they started out ½ meters away? Your calculation looks like this:

$$F = (G \, m_1 m_2) \, / \, r^2 = [(6.67 \times 10^{-11})(75)(75)] \, / \, \tfrac{1}{2}^2 = 1.5 \times 10^{-6} N$$

The force of attraction is roughly five millionths of an ounce. Maybe not enough to shake the surface of the Earth, but that's okay. The Earth's surface has its own forces to deal with.

Deriving the force of gravity on the Earth's surface

The equation for the force of gravity — $F = (G \, m_1 m_2) \, / \, r^2$ — holds true no matter how far apart two masses are. But you also come across a special gravitational case (which most of the work on gravity in this book is about) — the force of gravity on the surface of the Earth. Adding gravity to mass is where the difference between weight and mass comes in. Mass is considered a measure of an object's inertia, and its weight is the force exerted on it in a gravitational field. On the surface of the Earth, the two forces are related by the acceleration due to gravity: $F_g = mg$. Grams, kilograms, and slugs are all units of mass; dynes, newtons, and pounds are all units of weight.

Can you derive *g*, the acceleration due to gravity on the surface of the Earth, from Newton's law of gravitation? You sure can. The force on an object of mass m_1 near the surface of the Earth is

$$F = m_1 g$$

By Newton's third law (see Chapter 5), this force must also equal the following, where r_e is the radius of the Earth:

$$F = m_1 g = (G\ m_1 m_2)\ /\ r_e^2$$

The radius of the Earth, r_e, is about 6.38×10^6 meters, and the mass of the Earth is 5.98×10^{24} kg, so you have

$$F = m_1 g = (G\ m_1 m_2)\ /\ r_e^2 = [(6.67 \times 10^{-11})\ m_1\ (5.98 \times 10^{24})]\ /\ (6.38 \times 10^6)^2$$

Dividing both sides by m_1 gives you

$$g = [(6.67 \times 10^{-11})(5.98 \times 10^{24})]\ /\ (6.38 \times 10^6)^2 = 9.8\ \text{meters/second}^2$$

Newton's law of gravitation gives you the acceleration due to gravity on the surface of the Earth: 9.8 meters per second^2.

You can use Newton's law of gravitation to get the acceleration due to gravity, g, on the surface of the Earth just by knowing the gravitational constant G, the radius of the Earth, and the mass of the Earth. (Of course, you can measure g by letting an apple drop and timing it, but what fun is that when you can calculate it a roundabout way that requires you to first measure the mass of the Earth?)

Using the law of gravitation to examine circular orbits

In space, bodies are constantly orbiting other bodies due to gravity. Satellites (including the moon) orbit the Earth, the Earth orbits the Sun, the Sun orbits around the center of the Milky Way, the Milky Way orbits around the center of its local group of galaxies. This is big time stuff. In the case of orbital motion, gravity supplies the centripetal force that pushes the orbits. The force is quite a bit different than small-time orbital motion — like when you have a ball on a string — because for a given distance and two masses, the gravitational force is always going to be the same. You can't increase it to increase the speed of an orbiting planet as you can with a ball. The following sections examine the speed and period of orbiting bodies and Kepler's laws for orbiting bodies.

Calculating a satellite's speed

A particular satellite can have only one speed when in orbit around a particular body at a given distance because the force of gravity doesn't change. So, what's that speed? You can calculate it with the equations for centripetal force and gravitational force. You know that for a satellite of a particular mass, m_1, to orbit, you need a corresponding centripetal force (see section "Pulling Toward the Center: Centripetal Force"):

$$F_c = (m_1 v^2) / r$$

This centripetal force has to come from the force of gravity, so

$$F = (G \, m_1 m_2) / r^2 = (m_1 v^2) / r$$

You can rearrange this equation to get the speed:

$$v = \sqrt{\frac{Gm_2}{r}}$$

This equation represents the speed a satellite at a given radius must have in order to orbit if the orbit is due to gravity. The speed can't vary as long as the satellite has a constant orbital radius — that is, as long as it's going around in circles. This equation holds for any orbiting object where the attraction is the force of gravity, whether it's a man-made satellite orbiting the Earth or the Earth orbiting the Sun. If you want to find the speed for satellites that orbit the Earth, for example, you use the mass of the Earth in the equation:

$$v = \sqrt{\frac{Gm_E}{r}}$$

Man-made satellites typically orbit at heights of 400 miles from the surface of the Earth (about 640 kilometers or so). What's the speed of such a satellite? All you have to do is put in the numbers:

$$v = \sqrt{\frac{Gm_E}{r}} = \sqrt{\frac{(6.67 \times 10^{-11})(5.98 \times 10^{24})}{7.02 \times 10^6}} = 7.53 \times 10^3 \text{ meters per second}$$

This converts to about 16,800 miles per hour.

A few details you should note after reviewing the speed equation:

✔ The number you divide by in the previous equation, 7.02×10^6 meters, isn't the distance the satellite orbits above the Earth's surface, which is 640 kilometers. This is because you have to use the distance from the center of the Earth when you calculate gravitational attraction. Therefore, the distance you use in the equation is the distance between the two orbiting bodies. In this case, you add the distance from the center of the Earth to the surface of the Earth, 6.38×10^6 meters, to the 640 kilometers.

✔ The equation assumes that the satellite is high enough off the ground that it orbits out of the atmosphere. That assumption isn't really true, even at 400 miles, so satellites like this do feel air friction. Gradually, the drag of friction brings them lower and lower, and when they hit the atmosphere, they burn up on re-entry. When a satellite is less than 100 miles above the surface, its orbit decays appreciably each time it circles the Earth. (Look out below!)

✔ The equation is independent of mass. If the moon orbited at 640 kilometers rather than the man-made satellite, and you could ignore air friction and collisions with the Earth, it would have to go at the same speed as the satellite in order to preserve its close orbit (which would make for some pretty spectacular moonrises).

You can think of a satellite in motion over the Earth as always falling. The only thing that keeps it from falling straight to Earth is that its velocity points over the horizon. The satellite is falling, which is why astronauts in a space station are weightless, but its velocity takes it over the horizon — that is, over the curve of the world as it falls — so it doesn't get any closer to the Earth.

Sometimes, it's more important to know the period of an orbit rather than the speed, such as when you're counting on a satellite to come over the horizon before communication can take place.

Calculating the period of a satellite

The *period* of a satellite is the time it takes it to make one full orbit around an object. If you know the satellite's speed and the radius it orbits at (see the previous section), you can figure out its period. The satellite travels around the entire circumference of the circle — which is $2\pi r$ if r is the radius of the orbit — in the period, T. This means the orbital speed must be $2\pi r / T$, giving you

$$v = \sqrt{\frac{Gm_E}{r}} = \frac{2\pi r}{T}$$

If you solve this for the period of the satellite, you get

$$T = 2\pi \sqrt{\frac{r^3}{Gm_E}}$$

You, the intuitive physicist, may be wondering: What if you want to examine a satellite that simply stays stationary over the same place on the Earth at all times? In other words, a satellite whose period is the same as the Earth's 24-hour period? Can you do it? As you may know, such satellites do exist. They're very popular for communications, because they're always orbiting in the same spot; they don't disappear over the horizon and then reappear later. They also allow for global positioning satellites, or GPS. In cases of stationary satellites, the period, T, is 24 hours, or 86,400 seconds. Can you find the radius a stationary satellite needs to have? Using the equation for periods, you see that:

$$r^3 = (T^2 Gm_E) / 4\pi^2$$

Understanding Kepler's laws of orbiting bodies

Johannes Kepler, a German national born in the Holy Roman Empire, came up with three laws that help explain a great deal about orbits before Newton came up with his law of universal gravitation. Here are Kepler's laws:

✔ **Law 1:** Planets orbit in ellipses. (Note that a circle is also an ellipse.)

✔ **Law 2:** Planets move so that a line between the Sun and the planet sweeps out the same area in the same time, independent of where they are in their orbits.

✔ **Law 3:** The square of a planet's orbital period is proportional to its distance from the Sun cubed.

You can see how the third law was derived from Newton's laws by looking at the section "Calculating the period of a satellite." It takes the form of this equation: $r^3 = (T^2 G\, m_E) / 4\pi^2$. Although Kepler's third law says that T^2 is proportional to r^3, you can get the exact constant relating these quantities by using Newton's law of gravitation.

Plugging in the numbers, you get

$$r^3 = (T^2 Gm_E) / 4\pi^2 = [(8.64 \times 10^4)^2 (6.67 \times 10^{-11})(5.98 \times 10^{24})] / 4\pi^2 = 4.23 \times 10^7 \text{ meters}$$

You get a radius of 4.23×10^7 meters. Subtracting the Earth's radius of 6.38×10^6 meters, you get 3.59×10^7 meters, which converts to about 22,300 miles. This is the distance from the Earth geosynchronous satellites need to orbit. At this distance, they orbit the Earth at the same speed the Earth is turning, which means that they stay put over the same piece of real estate.

In practice, it's very hard to get the speed just right, which is why geosynchronous satellites have either gas boosters that can be used for fine-tuning or magnetic coils that allow them to move by pushing against the Earth's magnetic field.

Looping the Loop: Vertical Circular Motion

Maybe you've watched extreme sports on television and wondered how bikers or skateboarders can ride into a loop on a track and go upside down at the top of the loop without falling to the ground. Shouldn't gravity bring them down? How fast do they have to go? The answers to these vertical circular motion questions lie in centripetal force and the force of gravity.

Take a look at Figure 7-4, where a ball is looping around a track resembling a rollercoaster loop. A question you may come across in introductory physics classes asks, "What speed is necessary so that the ball makes the loop safely?" The crucial point is at the very top of the track — if the ball is going to peel away from its circular track, the top is where it will fall. To answer the crucial question, you must know what criterion the ball must meet to hold on. Ask yourself: "What's the constraint that the ball must meet?"

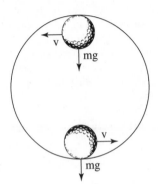

Figure 7-4:
The force and velocity of a ball on a circular track.

To travel in a loop, an object must have a net force acting on it that equals the centripetal force it needs to keep traveling in a circle. At the top of its path, as you can see in Figure 7-4, the ball barely stays in contact with the track. Other points along the track will provide normal force (see Chapter 6) because of the speed and the fact that the track is curved. If you want to find out what minimum speed an object needs to have to stay on a loop, you need to look at where the object is just barely in contact with the track — in other words, on the verge of falling out of its circular path.

The normal force the track applies to an object at the top is just about zero. The *only* force keeping the object on its circular track is the force of gravity, which means that at the apex, the speed of the object has to match the centripetal force provided by gravity to keep it going in a circle whose radius is the same as the radius of the loop. That means that if this is the centripetal force needed:

$$F_c = (mv^2) / r$$

The force of gravity at the top of the loop is

$$F_g = mg$$

And because F_g must equal F_c,

$$(mv^2) / r = mg$$

You can simplify this equation into the following form:

$$v = \sqrt{rg}$$

The mass of any object traveling around a circular track, such as a motorcycle or a racecar, drops out.

The square root of r times g is the minimum speed an object needs at the top of the loop in order to keep going in a circle. Any object with a slower speed will peel off the track at the top of the loop (it may drop back into the loop, but it won't be following the circular track at that point). For a practical example, if the loop from Figure 7-4 has a radius of 20 meters, how fast does the ball have to travel at the top of the loop in order to stay in contact with the track? Just put in the numbers:

$$v = \sqrt{rg} = \sqrt{(20.0)(9.8)} = 14.0 \text{ meters per second}$$

The golf ball has to travel 14.0 meters per second at the top of the track, which is about 31 miles per hour.

What if you want to do the same trick on a flaming circular loop on a motorcycle to impress your pals? The same speed applies — you need to be going about 31 miles per hour minimum at the top of the track, which has a radius of 20 meters. If you want to try this at home, don't forget that this is the speed you need at the *top* of the track — you have to go faster at the bottom of the track in order to travel at 31 miles per hour at the top, simply because you're twice the radius, or 40 meters, higher up in the air, much like having coasted to the top of a 40-meter hill.

So, how much faster do you need to go at the bottom of the track? Check out Part III of this book, where kinetic energy (the kind of energy moving motorcycles have) is turned into potential energy (the kind of energy that motorcycles have when they're high up in the air against the force of gravity).

Part III

Manifesting the Energy to Work

In this part . . .

*I*f you drive a car up a hill and park it, it still has energy — potential energy. If the brake slips and the car rolls down the hill, it has a different kind of energy at the bottom — kinetic energy. This part alerts you to what energy is all about and how the work you put into moving and stretching objects becomes energy. Thinking in terms of work and energy allows you to solve problems that Newton's laws don't even let you attempt.

Chapter 8

Getting Some Work out of Physics

*Y*ou know all about work; it's what you do when you have to do physics problems. You sit down with your calculator, you sweat a little, and you get through it. You've done your work. Unfortunately, that doesn't count for work in physics terms.

You do work in physics by multiplying a force by the distance over which it acts. That may not be your boss's idea of work, but it gets the job done in physics. Along with the basics of work, I use this chapter to introduce kinetic and potential energy, look at conservative and nonconservative forces, and examine mechanical energy and power. Time to get to work.

Work: It Isn't What You Think

Work is defined as an applied force over a certain distance. In physics jargon, you do work by applying a constant force, F, over a distance, s, where the angle between F and s is θ and is equal to Fs cos θ. In layman's terms, if you push a 1,000-pound hockey puck for some distance, physics says that the work you do is the component of the force you apply in the direction of travel multiplied by the distance you go.

To get a picture of the full work spectrum, you need to look across different systems of measurement. After you have the measurement units down, you can look at practical working examples, such as pushing and dragging.

Working on measurement systems

Work is a scalar, not a vector (meaning it only has a magnitude, not a direction; more on scalars and vectors in Chapter 4). Because work is force times distance, Fs cos θ, it has the units Newton-meter in the meters-kilograms-seconds (MKS) system (see Chapter 2). Newton-meters are awkward to work with, so they have a special name — *Joules.* For conversion purposes, 1 Newton multiplied by 1 meter = 1 Joule, or 1J.

In the centimeter-gram-second (CGS) system, Fs cos θ has the units dyne-centimeter, which is also called the *erg* (a great name for a unit of work because it sounds like what you would say when pushing a heavy weight). For conversion purposes, 1 erg = 1 dyne-centimeter. What about the foot-pound-second system? In this system, work has the units foot-pound.

Pushing your weight

Holding heavy objects — like, say, a set of exercise weights — up in the air seems to take a lot of work. In physics terms, however, that isn't true. Even though holding up weights may take a lot of biological work, no physics work takes place if the weights aren't moving. Plenty of chemistry happens as your body supplies energy to your muscles, and you may feel a strain, but if you don't move anything, you don't do work in physics terms.

Motion is a requirement of work. For example, say that you're pushing a huge gold ingot home after you explore a cave down the street, as shown in Figure 8-1. How much work do you have to do to get it home? First, you need to find out how much force pushing the ingot requires.

Figure 8-1:
Pushing
requires
plenty of
work in
physics
terms when
the object is
in motion.

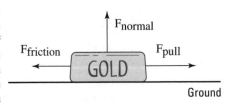

The kinetic coefficient of friction (see Chapter 6), μ_k, between the ingot and the ground is 0.25, and the ingot has a mass of 1,000 kg. What's the force you have to exert to keep the ingot moving without accelerating it? Start with this equation from Chapter 6:

$$F_F = \mu_k F_N$$

Assuming that the road is flat, the magnitude of the normal force, F_N, is just *mg* (mass times gravity). That means that

$$F_F = \mu_k F_N = \mu_k\, mg$$

where *m* is the mass of the ingot and *g* is the acceleration due to gravity on the surface of the Earth. Plugging in the numbers gives you

$$F_F = \mu_k F_N = \mu_k\, mg = (0.25)(1{,}000 \text{ kg})(9.8 \text{ meters per second}^2) = 2{,}450\text{N}$$

You have to apply a force of 2,450 Newtons to keep the ingot moving without accelerating. Say that your house is 3 kilometers away, or 3,000 meters. To get the ingot home, you have to do this much work:

$$W = Fs \cos \theta$$

Because you're pushing the ingot, the angle between F and s is 0°, and $\cos \theta = 1$, so plugging in the numbers gives you

$$W = Fs \cos \theta = (2{,}450)(3{,}000)(1) = 7.35 \times 10^6\text{J}$$

You need to do 7.35×10^6J of work to move your ingot home. Want some perspective? Well, to push 1 kilogram 1 meter, you have to supply a force of 9.8N (about 2.2 pounds) over that distance, which takes 9.8J of work. To get your ingot home, you need 750,000 times that. Put another way, 1 kilocalorie equals 4,186J. A kilocalorie is commonly called a Calorie (capital C) in nutrition, which you see on candy bar labels, so to move the ingot home, you need to expend about 1,755 calories. Time to get out the energy bars!

You can also do this conversion in terms of kilowatt-hours, which you may be familiar with from electric bills. One kilowatt-hour (kWh) = 3.6×10^6J, so you need about 2.0 kilowatt-hours of work to get the ingot home.

Taking a drag

If you have a backward-type personality, you may prefer to drag objects rather than push them. It may be easier to drag heavy objects, especially if you can use a tow rope, as shown in Figure 8-2. When you're pulling at an angle θ, you're not applying a force in the same direction as the direction of motion. To find the work in this case, all you have to do is find the component of the force along the direction of travel. Work properly defined is the force along the direction of travel multiplied by the distance traveled:

$$W = F_{pull}s \cos \theta$$

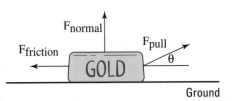

Figure 8-2:
Dragging an
object
requires
more force
due to the
angle.

Assume that the angle at which you're pulling is small, so you're not lifting the ingot (which would lessen the normal force and therefore the friction). You need a force of 2,450N along the direction of travel to keep the object in motion (see the previous section for the calculation), which means that you have to supply a force of

$$F_{pull} \cos \theta = 2,450N$$

Therefore,

$$F_{pull} = 2,450N / \cos \theta$$

If θ = 10°, you have to supply a force of

$$F_{pull} = 2,450N / \cos 10° = 2,490N$$

Because only the work along the direction of travel counts, and because you're actually pulling on the tow rope at an angle of 10°, you need to provide more force to get the same amount of work done, assuming the object travels the same path to your house as the ingot in Figure 8-1.

Considering Negative Work

You've just gone out and bought the biggest television your house can handle. You finally get the TV home, and you have to lift it up the porch stairs. It's a heavy one — about 100 kg, or 220 pounds — and as you lift it up the first stair — a distance of about ½ meters — you think you should have gotten some help because of how much work you're doing (***Note:*** F equals mass times acceleration, or 100 times *g*, the acceleration due to gravity; and θ is 0° because you're lifting upward, the direction the TV is moving):

$$W_1 = Fs \cos \theta = (100)(9.8)(\frac{1}{2})(1.0) = 490J$$

However, as you get the TV to the top of the step, your back decides that you're carrying too much weight and advises you to drop it. Slowly, you let it fall back to its original position and take a breather. How much work did you do on the way down? Believe it or not, you did *negative* work on the TV, because the force you applied (upward) was in the opposite direction of travel (downward). In this case, θ = 180°, and cos θ = –1. This is what you get when you solve for the work:

$$W_2 = Fs \cos \theta = (100)(9.8)(\frac{1}{2})(-1) = -490J$$

The net work you've done is $W = W_1 + W_2 = 0$, or zero work. That makes sense, because the TV is right back where it started.

If the force moving the object has a component in the same direction as the motion, the work that force does on the object is positive. If the force moving the object has a component in the opposite direction of the motion, the work done on the object is negative.

Getting the Payoff: Kinetic Energy

When you start pushing or pulling an object with a constant force, it starts to move if the force you exert is greater than the net forces resisting you (such as friction and gravity). And if the object starts to move at some speed, it will acquire *kinetic energy*. Kinetic energy is the energy an object has because of its motion. Energy is the ability to do work.

For example, say you come to a particularly difficult hole of miniature golf, where you have to hit the ball through a loop. The golf ball enters the loop with a particular speed; in physics terms, it has a certain amount of kinetic energy. Assume that when it gets to the top of the loop, it stops dead. This means it no longer has any kinetic energy at all. However, now that it's at the top of the loop, it sits higher than it was before, and when it drops back down — assuming it stays on the track and there was no friction — it will have the same speed when it gets to the bottom of the track as it had when it first entered the track.

If the golf ball has 20J of kinetic energy at the bottom of the loop, the energy is due to its motion. At the top of the loop, it had no motion, so it had no kinetic energy. However, it took some work to get the golf ball to the top, and that work was 20J, so the golf ball at rest has 20J of what's called *potential energy*. The golf ball has potential energy because if it falls, that 20J of energy will be available; if it falls and stays on the track, that 20J of potential energy, which it had because of its height, will become 20J of kinetic energy again. (For more on potential energy, see the section "Energy in the Bank: Potential Energy" later in this chapter.)

At the bottom of the loop, the golf ball has 20J of kinetic energy and is moving; at the top of the loop, it has 20J of potential energy and isn't moving; and when it comes back down, it has 20J of kinetic energy again, as shown in Figure 8-3. The golf ball's total energy stays the same — 20J at the bottom of the loop, 20J at the top of the loop. The energy takes different forms — kinetic when it's moving and potential when it isn't moving but is higher up — but it's the same. In fact, the golf ball's energy is the same at any point around the loop, and physicists who've measured this kind of phenomenon call this the principle of *conservation of mechanical energy*. I discuss this later in this chapter in the section "Up, Down, and All Around: The Conversion of Mechanical Energy," so stay tuned.

Figure 8-3:
An object circling a loop without friction has the same energy throughout; it just takes different forms.

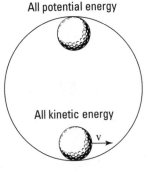

All potential energy

All kinetic energy

v

Where does the kinetic energy go when friction is involved? If a block is sliding along a horizontal surface and there's friction, the block goes more and more slowly until it comes to a stop. The kinetic energy goes away, and you see no increase in potential energy. What happened? The block's kinetic energy dissipated as heat. Friction heated both the block and the surface.

You know the ins and outs of kinetic energy. So how do you calculate it?

Breaking down the kinetic energy equation

The work that you put into accelerating an object — that is, into its motion — becomes the object's kinetic energy, KE. The equation to find KE is

$$KE = \tfrac{1}{2} m v_f^2$$

Given a mass m going at a speed v, you can calculate an object's kinetic energy. Say, for example, that you apply a force to a model airplane in order to get it flying and that the plane is accelerating. Here's the equation for force:

$$F = ma$$

You know that force equals mass times acceleration, and you know from the previous sections in this chapter that the work done on the plane, which becomes its kinetic energy, equals the following:

$$W = Fs \cos \theta$$

Assume that you're pushing in the same direction that the plane is going; in this case, $\cos \theta = 1$, and you find that

$$W = Fs = mas$$

You can tie this equation to the final and original velocity of the object (see Chapter 3 for that equation) to find a:

$$v_f^2 - v_o^2 = 2as$$

where v_f equals final velocity and v_o equals initial velocity. In other words,

$$a = (v_f^2 - v_o^2) / 2s$$

Plugging in for a in the equation for work, $W = mas$, you get:

$$W = \tfrac{1}{2} m(v_f^2 - v_o^2)$$

If the initial velocity is zero, you get

$$W = \tfrac{1}{2} m v_f^2$$

This is the work that you put into accelerating the model plane — that is, into the plane's motion — and that work becomes the plane's kinetic energy, KE:

$$KE = \tfrac{1}{2} m v_f^2$$

Putting the kinetic energy equation to use

You normally use the kinetic energy equation to find the kinetic energy of an object when you know its mass and velocity. Say, for example, that you're at a pistol-firing range, and you fire a 10-gram bullet with a velocity of 600 meters per second at a target. What's the bullet's kinetic energy? The equation to find kinetic energy is

$$KE = \frac{1}{2} mv^2$$

All you have to do is plug in the numbers, remembering to convert from grams to kilograms first to keep the system of units consistent throughout the equation:

$$KE = \frac{1}{2} mv^2 = \frac{1}{2} (0.01)(600^2) = 1{,}800J$$

The bullet has 1,800 Joules of energy, which is a lot of Joules to pack into a 10-gram bullet. However, you can also use the kinetic energy equation if you know how much work goes into accelerating an object and you want to find, say, its final speed. For example, say you're on a space station, and you have a big contract from NASA to place satellites in orbit. You open the station's bay doors and grab your first satellite, which has a mass of 1,000 kg. With a tremendous effort, you hurl it into its orbit, using a force of 2,000N, applied in the direction of motion, over 1 meter. What speed does the satellite attain relative to the space station? The work you do is equal to

$$W = Fs \cos \theta$$

Because $\theta = 0°$ here (you're pushing the satellite straight on), $W = Fs$:

$$W = Fs = (2{,}000)(1.0) = 2{,}000J$$

Your work goes into the kinetic energy of the satellite, so

$$W = Fs = (2{,}000)(1.0) = 2{,}000J = \frac{1}{2} mv^2$$

From here, you can figure the speed by putting v on one side, because m equals 1000 kg and W equals 2000 Joules:

$$v = \sqrt{\frac{(2)(2000)}{1000}} = 2 \text{ meters per second}$$

The satellite ends up with a speed of 2 meters per second relative to you — enough to get it away from the space station and into its own orbit.

Bear in mind that forces can also do negative work. If you want to catch a satellite and slow it to 1 meter per second with respect to you, the force you apply to the satellite is in the opposite direction of its motion. That means it loses kinetic energy, so you did negative work on it.

You only have to worry about one force in this example — the force you apply to the satellite as you launch it. But in everyday life, multiple forces act on an object, and you have to take them into account.

Calculating kinetic energy by using net force

If you want to find the total work on an object and convert that into its kinetic energy, you have to consider only the work done by the net force. In other words, you convert only the net force into kinetic energy. Other forces may be acting, but opposing forces, such as a normal force and the force of gravity (see Chapter 6), cancel each other out. For instance, when you play tug-of-war against your equally strong friends, you pull against each other and nothing moves. You have no net increase in kinetic energy from the two forces.

For example, take a look at Figure 8-4. You may want to determine the speed of the 100-kg refrigerator at the bottom of the ramp, using the fact that the work done on the refrigerator goes into its kinetic energy. How do you do that? You start by determining the net force on the refrigerator and then finding out how much work that force does. Converting that net-force work into kinetic energy lets you calculate what the refrigerator's speed will be at the bottom of the ramp.

Figure 8-4:
You find the net force acting on an object to find its speed at the bottom of a ramp.

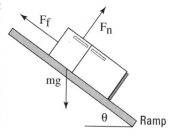

What's the net force acting on the refrigerator? In Chapter 6, you find that the component of the refrigerator's weight acting along the ramp is

$$F_g \text{ ramp} = mg \cos(90° - \theta) = mg \sin \theta$$

where m is the mass of the refrigerator and g is the acceleration due to gravity. The normal force (see Chapter 6) is

$$F_N = mg \sin (90° - \theta) = mg \cos \theta$$

which means that the kinetic force of friction (see Chapter 6) is

$$F_F = \mu_k F_N = \mu_k mg \sin (90° - \theta) = \mu_k mg \cos \theta$$

where μ_k is the kinetic coefficient of friction. The net force accelerating the refrigerator down the ramp, F_{net}, therefore, is

$$F_{net} = F_g \text{ ramp} - F_F = mg \sin \theta - \mu_k mg \cos \theta$$

You're most of the way there! If the ramp is at a 30° angle to the ground and has a kinetic coefficient of friction of 0.15, plugging the numbers into this equation results in the following:

$$F_{net} = (100)(9.8)(\sin 30°) - (0.15)(100)(9.8)(\cos 30°) = 490N - 127N = 363N$$

The net force acting on the refrigerator is 363N. This net force acts over the entire 3.0-meter ramp, so the work done by this force is

$$W = F_{net\ s} = (363)(3.0) = 1,089J$$

You find that 1,089J goes into the refrigerator's kinetic energy. That means you can find the refrigerator's kinetic energy like this:

$$W = 1,089J = KE = \frac{1}{2} mv^2$$

You want the speed here, so solving for v gives you

$$v = \sqrt{\frac{(2)(1089)}{100}} = 4.67 \text{ meters per second}$$

The refrigerator will be going 4.67 meters per second at the bottom of the ramp.

Energy in the Bank: Potential Energy

There's more to motion than kinetic energy — an object can also have *potential energy*, which is the energy it has because of its position, or stored energy. The energy is called potential because it can be converted back to kinetic or other forms of energy at any time.

Say, for example, you have the job of taking your little cousin Jackie to the park, and you put the little tyke on the slide. Jackie starts at rest and then accelerates, ending up with quite a bit of speed at the bottom of the slide. You sense physics at work here. Taking out your clipboard, you put Jackie higher up the slide and let go, watching carefully. Sure enough, Jackie ends up going even faster at the bottom of the slide. You decide to move Jackie even higher up. Suddenly, Jackie's mother shows up and grabs him from you. That's enough physics for one day.

What was happening on the slide? Where did Jackie's kinetic energy come from? It came from the work you did lifting Jackie against the force of gravity. Jackie sits at rest at the bottom of the slide, so he has no kinetic energy. If you lift him to the top of the slide and hold him, he waits for the new trip down the slide, so he has no motion and no kinetic energy. However, you did work lifting him up against the force of gravity, so he has potential energy. As Jackie slides down the (frictionless) slide, gravity turns your work, and the potential energy you create, into kinetic energy.

Working against gravity

How much work do you do when you lift an object against the force of gravity? Say, for example, that you want to store a cannonball on an upper shelf at height h above where the cannonball is now. The work you do is

$$W = Fs \cos \theta$$

In this case, F equals force, s equals distance, and θ is the angle between them. The force on an object is mg (mass times the acceleration due to gravity, 9.8 meters per second^2), and when you lift the cannonball straight up, $\theta = 0°$, so

$$W = Fs \cos \theta = mgh$$

The variable h here is the distance you lift the cannonball. To lift the ball, you have to do a certain amount of work, or m times g times h. The cannonball is stationary when you put it on the shelf, so it has no kinetic energy. However, it does have potential energy, which is the work you put into the ball to lift it to its present position.

If the cannonball rolls to the edge of the shelf and falls off, how much kinetic energy would it have just before it strikes the ground (which is where it started when you first lifted it)? It would have mgh joules of kinetic energy at that point. The ball's potential energy, which came from the work you put in lifting it, converts to kinetic energy thanks to the fall.

In general, you can say that if you have an object of mass m near the surface of the Earth where the acceleration due to gravity is g, at a height h, the potential energy of that mass compared to what it would be if it were at height 0 is

$$PE = mgh$$

And if you move an object vertically against the force of gravity from height h_o to height h_f, its change in potential energy is

$$\Delta PE = mg\,(h_f - h_o)$$

The work you perform on the object changes its potential energy.

Converting potential energy into kinetic energy

Objects can have different kinds of potential energy — all you need to do is perform work on an object against a force, such as when an object is connected to a spring and you pull the spring back. However, gravity is a very common source of potential energy in physics problems. Gravitational potential energy for a mass m at height h near the surface of the Earth is mgh more than it would be at height 0. (It's up to you where you choose height 0.)

For example, say that you lift a 40-kg cannonball onto a shelf 3.0 meters from the floor, and the ball rolls and slips off, headed toward your toes. If you know the potential energy involved, you can figure out how fast the ball will be going when it reaches the tips of your shoes. Resting on the shelf, the cannonball has this much potential energy with respect to the floor:

$$PE = mgh = (40 \text{ kg})(9.8 \text{ meters per second}^2)(3.0 \text{ meters}) = 1{,}176J$$

The cannonball has 1,176 joules of potential energy stored by virtue of its position in a gravitational field. What happens when it drops, just before it touches your toes? That potential energy is converted into kinetic energy. So, how fast will the cannonball be going at toe impact? Because its potential energy is converted into kinetic energy, you can write the problem as the following (see the section "Getting the Payoff: Kinetic Energy" earlier in this chapter for an explanation of the kinetic energy equation):

$$PE = mgh = (40)(9.8)(3.0) = 1{,}176J = KE = \tfrac{1}{2}\,mv^2$$

Plugging in the numbers and putting velocity on one side, you get the speed:

$$v = \sqrt{\frac{(2)(1176)}{40}} = 7.67 \text{ meters/second}$$

The velocity of 7.67 meters per second converts to about 25 feet per second. You have a 40-kg cannonball — or about 88 pounds — dropping onto your toes at 25 feet per second. You play around with the numbers and decide you don't like the results. Prudently, you turn off your calculator and move your feet out of the way.

Choose Your Path: Conservative versus Nonconservative Forces

The work a *conservative force* does on an object is path-independent; the actual path taken by the object makes no difference. Fifty meters up in the air has the same gravitational potential energy whether you get there by taking the steps or by hopping on a Ferris wheel. That's different from the force of friction, for example, which dissipates kinetic energy as heat. When friction is involved, the path you take does matter — a longer path will dissipate more kinetic energy than a short one. For that reason, friction isn't a conservative force; it's a *nonconservative force*.

For example, you and some buddies arrive at Mt. Newton, a majestic peak that soars h meters into the air. You can take two ways up — the quick way or the scenic route. Your friends drive up the quick route, and you drive up the scenic way, taking time out to have a picnic and to solve a few physics problems. They greet you at the top by saying, "Guess what? Our potential energy compared to before is *mgh* greater."

"Me too," you say, looking out over the view. Note this equation (originally presented in the section "Working against gravity," earlier in this chapter):

$$\Delta PE = mg \left(h_f - h_o \right)$$

This equation basically states that the actual path you take when going vertically from h_o to h_f doesn't matter. All that matters is your beginning height compared to your ending height. Because the path taken by the object against gravity doesn't matter, gravity is a conservative force.

Here's another way of looking at conservative and nonconservative forces. Say that you're vacationing in the Alps and that your hotel is at the top of Mt. Newton. You spend the whole day driving around — down to a lake one minute, to the top of a higher peak the next. At the end of the day, you end up back at the same location: your hotel on top of Mt. Newton.

What's the change in your gravitational potential energy? In other words, how much net work did gravity perform on you during the day? Because gravity is a conservative force, the change in your gravitational potential energy is 0. Because you've experienced no net change in your gravitational potential energy, gravity did no net work on you during the day.

The road exerted a normal force on your car as you drove around (see Chapter 6), but that force was always perpendicular to the road, so it didn't do any work.

Conservative forces are easier to work with in physics because they don't "leak" energy as you move around a path — if you end up in the same place, you have the same amount of energy. If you have to deal with forces like friction, including air friction, the situation is different. If you're dragging something over a field carpeted with sandpaper, for example, the force of friction does different amounts of work on you depending on your path. A path that's twice as long will involve twice as much work overcoming friction. The work done depends on the path you take, which is why friction is a nonconservative force.

Allow me to qualify the idea that friction is a nonconservative force. What's really not being conserved around a track with friction is the total potential and kinetic energy, which taken together is *mechanical energy*. When friction is involved, the loss in mechanical energy goes into heat energy. In a way, you could say that the total amount of energy doesn't change if you include that heat energy. However, the heat energy dissipates into the environment quickly, so it isn't recoverable or convertible. For that and other reasons, physics often works in terms of mechanical energy.

Up, Down, and All Around: The Conservation of Mechanical Energy

Mechanical energy is the sum of potential and kinetic energy, or the energy acquired by an object upon which work is done. *The conservation of mechanical energy,* which occurs in the absence of nonconservative forces, makes your life much easier when it comes to solving physics problems. Say, for example, that you see a rollercoaster at two different points on a track — point 1 and point 2 — so that the coaster is at two different heights and two different speeds at those points. Because mechanical energy is the sum of the potential (mass × gravity × height) and kinetic (½[mass × velocity^2]) energies, at point 1, the total mechanical energy is

$$W_1 = mgh_1 + \tfrac{1}{2} mv_1^2$$

At point 2, the total mechanical energy is

$$W_2 = mgh_2 + \tfrac{1}{2} mv_2^2$$

What's the difference between W_2 and W_1? If friction is present, for example, or any other nonconservative forces, the difference is equal to the net work the nonconservative forces do, W_{nc} (see the previous section for an explanation of net work):

$$W_2 - W_1 = W_{nc}$$

On the other hand, if nonconservative forces perform no net work, $W_{nc} = 0$, which means that

$$W_2 = W_1$$

or

$$mgh_1 + \tfrac{1}{2}\,mv_1{}^2 = mgh_2 + \tfrac{1}{2}\,mv_2{}^2$$

These equations represent the *principle of conservation of mechanical energy.* The principle says that if the net work done by nonconservative forces is zero, the total mechanical energy of an object is conserved; that is, it doesn't change.

Another way of rattling off the principle of conservation of mechanical energy is that at point 1 and point 2,

$$PE_1 + KE_1 = PE_2 + KE_2$$

You can simplify that mouthful to the following:

$$E_1 = E_2$$

where E is the total mechanical energy at any one point. In other words, an object always has the same amount of energy as long as the net work done by nonconservative forces is zero.

Note that you can cancel out the mass, *m*, in the previous equation, which means that if you know three of the values, you can solve for the fourth:

$$gh_1 + \tfrac{1}{2}\,v_1{}^2 = gh_2 + \tfrac{1}{2}\,v_2{}^2$$

Breaking apart the equation for mechanical energy allows you to solve for individual variables, such as velocity and height.

Determining final velocity with mechanical energy

"Serving as a roller coaster test pilot is a tough gig," you say as you strap yourself into the Physics Park's new Bullet Blaster III coaster. "But someone has to do it." The crew closes the hatch and you're off down the 400-meter, totally frictionless track. Halfway down, however, the speedometer breaks. How can you record your top speed when you get to the bottom? No problem; all you need is the principle of conservation of mechanical energy, which says that if the net work done by nonconservative forces is zero, the total mechanical energy of an object is conserved. From the previous section, you know that

$$mgh_1 + \frac{1}{2}mv_1^2 = mgh_2 + \frac{1}{2}mv_2^2$$

You can make this equation a little easier. Your initial velocity is 0, your final height is 0, and you can divide both sides by m, so you get

$$gh_1 = \frac{1}{2}v_2^2$$

Much nicer. Solving for v_2 gives you

$$\sqrt{(9.8)(400)(2)} = 89 \text{ meters per second}$$

The coaster travels at 89 meters per second, or about 198 miles per hour, at the bottom of the track. Should be fast enough for most kids.

Determining final height with mechanical energy

Besides determining variables such as final speed with the principle of conservation of mechanical energy, you can determine final height if you want to. At this very moment, for example, Tarzan is swinging on a vine over a crocodile-infested river at a speed of 13.0 meters per second. He needs to reach the opposite river bank 9.0 meters above his present position in order to be safe. Can he swing it? The principle of conservation of mechanical energy gives you the answer:

$$mgh_1 + \frac{1}{2}mv_1^2 = mgh_2 + \frac{1}{2}mv_2^2$$

At Tarzan's maximum height at the end of the swing, his speed, v_2, will be 0, and assuming $h_1 = 0$, you can relate h_2 to v_1 like this:

$$\frac{1}{2}v_1^2 = gh_2$$

This means that:

$$h_2 = v_1^2 / 2g = 13.0^2 / [(2)(9.8)] = 8.6 \text{ meters}$$

Tarzan will come up 0.4 meters short of the 9.0 meters he needs to be safe, so he needs some help.

Powering Up: The Rate of Doing Work

Sometimes, it isn't just the amount of work you do but the rate at which you do work that's important, and rate is reflected in power. The concept of power gives you an idea of how much work you can expect in a certain amount of time. *Power* in physics is the amount of work done divided by the time it takes, or the *rate*. Here's what that looks like in equation form:

$$P = W / t$$

Assume you have two speedboats, for example, and you want to know which one will get you to 120 miles per hour faster. Ignoring silly details like friction, it will take you the same amount of work to get up to that speed, but what about how long it will take? If one boat takes three weeks to get you to 120, that may not be the one you take to the races. In other words, the amount of work you do in a certain amount of time can make a big difference.

If the work done at any one instant varies, this represents the average power over the entire time *t*. An average quantity in physics is often written with a bar over it, like the following for average power:

$$\bar{P} = W / t$$

Before you can jump into your power trip, however, it helps to know what units you're dealing with and the various ways to get to your answer.

Common units of power

Power is work divided by time, so power has the units of Joules per second, which is called the *watt* — a familiar term for just about anybody that uses anything electrical. You abbreviate a watt as simply W, so a 100-watt light bulb uses 100W of power.

You occasionally run across symbolization conflicts in physics, such as the W for watts and the W for work. This conflict isn't serious, however, because one symbol is for units (watts) and one is for a concept (work). Capitalization is standard, so be sure to pay attention to units versus concepts.

Note also that because work is a scalar quantity (see Chapter 4) — as is time — power is a scalar as well. Other units of power include the foot-pound per second (ft-lbs/s) and the horsepower (hp). One hp = 550 ft-lbs/s = 745.7 W.

Say, for example, that you're in a horse-drawn sleigh on the way to your grandmother's house. At one point, the horse accelerates the 500-kg sleigh from 1 meter per second to 2 meters per second in 2.0 seconds. What power does the move take? Assuming no friction on the snow, the total work done is

$$W = \tfrac{1}{2} mv_2^2 - \tfrac{1}{2} mv_1^2$$

Plugging in the numbers gives you

$$W = \tfrac{1}{2} mv_2^2 - \tfrac{1}{2} mv_1^2 = \tfrac{1}{2}(500)(2^2) - \tfrac{1}{2}(500)(1^2) = 1{,}000 - 250 = 750J$$

Because the horse does this work in 2 seconds, the power needed is

$$P = 750J \,/\, 2.0 \text{ seconds} = 375 \text{ watts}$$

One horsepower = 745.7 watts, so the horse moves at about one-half horsepower — not too bad for a one-horse open sleigh.

Alternate calculations of power

Because work equals force times distance, you can write the equation for power the following way, assuming that the force acts along the direction of travel:

$$P = W \,/\, t = Fs \,/\, t$$

where s is the distance traveled. However, the object's speed, v, is just s divided by t, so the equation breaks down to

$$P = W \,/\, t = Fs \,/\, t = Fv$$

That's an interesting result — power equals force times speed? Yep, that's what it says. However, because you often have to account for acceleration when you apply a force, you usually write the equation in terms of average power and average speed:

$$\bar{P} = F\bar{v}$$

Chapter 9

Putting Objects in Motion: Momentum and Impulse

*T*his chapter is all about the topics you need to know for all your travels: momentum and impulse. Both topics are very important to *kinematics,* or the study of objects in motion. After you have these topics under your belt, you can start talking about what happens when objects collide and go bang (hopefully not your car or bike). Sometimes they bounce off each other (like when you hit a tennis ball with a racket), and sometimes they stick together (like a dart hitting a dart board), but with the knowledge of impulse and momentum you pick up in this chapter, you can handle either case.

Looking at the Impact of Impulse

In physics terms, *impulse* tells you how much the momentum of an object will change when a force is applied for a certain amount of time (see the following section for a discussion on momentum). Say, for example, that you're shooting pool. Instinctively, you know how hard to tap each ball to get the results you want. The 9 ball in the corner pocket? No problem — tap it and there it goes. The 3 ball bouncing off the side cushion into the other corner pocket? Another tap, this time a little stronger.

The taps you apply are called *impulses*. Take a look at what happens on a microscopic scale, millisecond by millisecond, as you tap a pool ball. The force you apply with your cue appears in Figure 9-1 — the tip of each cue has a cushion, so the impact of the cue is spread out over a few milliseconds. The impact lasts from the time when the cue touches the ball, t_o, to the time when the ball loses contact with the cue, t_f. As you can see from Figure 9-1, the force exerted on the ball changes during that time; in fact, it changes drastically, and if you had to know what the force was doing at any one millisecond, it would be hard to figure out without some fancy equipment.

Figure 9-1: Examining force versus time gives you the impulse you apply on objects.

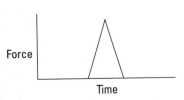

Because the pool ball doesn't come with any fancy equipment, you have to do what physicists normally do, which is to talk in terms of the average force over time: $\Delta t = t_f - t_o$. You can see what that average force looks like in Figure 9-2. Speaking as a physicist, you say that the impulse — or the tap — provided by the pool cue is the average force multiplied by the time that you apply the force, $\Delta t = t_f - t_o$. Here's the equation for impulse:

$$\text{Impulse} = \overline{F}\Delta t$$

Figure 9-2: The average force depends on the amount of time you apply the force.

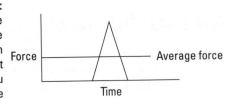

Note that this is a vector equation, meaning it deals with both direction and magnitude (see Chapter 4). Impulse is a vector, and it's in the same direction as the average force (which itself may be a net vector sum of other forces).

You get impulse by multiplying Newtons by seconds, so the units of impulse are Newton-seconds in the meter-kilogram-second (MKS) system (see Chapter 2), dyne-seconds in the centimeter-gram-second (CGS) system, and pound-seconds in the pound-foot-second system.

Gathering Momentum

When you apply an impulse to an object, the impulse can change its motion (see the previous section for more info on impulse). What does that mean? It means that you affect the object's momentum. Momentum is a concept most people have heard of — in physics terms, *momentum* is proportional to both mass and velocity, and to make your job easy, physics defines it as the product of mass times velocity. Momentum is a big concept both in introductory physics and in some advanced topics like high-energy particle physics, where the components of atoms zoom around at high speeds. When they collide, you can often predict what will happen based on your knowledge of momentum.

Even if you're unfamiliar with the physics of momentum, you're already familiar with the general idea. Catching a runaway car going down a steep hill is a problem because of its momentum. If a car without any brakes is speeding toward you at 40 miles per hour, it may not be a great idea to try to stop it simply by standing in its way and holding out your hand, unless you're Superman. The car has a lot of momentum, and bringing it to a stop requires plenty of effort. Same thing for an oil tanker that you need to bring to a stop. A load of oil sits in these tankers, and their engines aren't strong enough to make them turn or stop on a dime. Therefore, it can take an oil tanker 20 miles or more to come to a stop. All because of the ship's momentum.

Shouldn't *p* stand for *physics* rather than *momentum?*

Why does physics use the symbol *p* to represent momentum? It doesn't seem to fit (there's no "p" in "momentum"). I've never heard any reason for it, and no physics gurus I've asked have any idea either. The symbol is something people go along with, without having any idea why, like daylight savings time. Probably the best reason you can give is that all the other letters were taken when it came time to give one to momentum — you can't use *m* for momentum, for example, because *m* stands for mass.

The more mass that's moving (think of an oil tanker), the more momentum the mass has. The more velocity it has (think of an even faster oil tanker), the more momentum it has. The symbol for momentum is *p,* so you can say that

$$p = mv$$

Momentum is a vector quantity, meaning that it has a magnitude and a direction (see Chapter 4), and the magnitude is in the same direction as the velocity — all you have to do to get the momentum of an object is to multiply its mass by its velocity. Because you multiply mass by velocity, the units for momentum are kilograms-meters per second, kg-m/s, in the MKS system and grams-centimeters per second in the CGS system.

The Impulse-Momentum Theorem: Relating Impulse and Momentum

You can connect the impulse you give to an object — like striking a pool ball with a cue — with the object's change in momentum; all it takes is a little algebra and the process you explore in this section, called the *impulse-momentum theorem.* What makes the connection easy is that you can play with the equations for impulse and momentum (see the previous two sections) to simplify them so you can relate the two topics. What equations does physics have in its arsenal that connect these two? Relating force and velocity is a start. For example, force equals mass times acceleration (see Chapter 5), and the definition of average acceleration is

$$a = \Delta v / \Delta t = (v_f - v_o) / (t_f - t_o)$$

where v stands for velocity and t stands for time. Now you may realize that after you multiply that by the mass, you get force, which brings you closer to working with impulse:

$$F = ma = m(\Delta v / \Delta t) = m([v_f - v_o] / [t_f - t_o])$$

Now you have force in the equation. To get impulse, multiply that force by Δt, the time over which you apply the force:

$$F\Delta t = ma\Delta t = m(\Delta v / \Delta t)\Delta t = m\Delta v = m(v_f - v_o)$$

Take a look at the final term, $m(v_f - v_o)$. Because momentum equals *mv* (see the section "Gathering Momentum" earlier in this chapter), this is just the difference in the object's initial and final momentum: $p_f - p_o = \Delta p$. Therefore, you can add that to the equation:

$$F\Delta t = ma\Delta t = m\Delta v = m(v_f - v_o) = mv_f - mv_o = p_f - p_o = \Delta p$$

Now take a look at the term on the left, $F\Delta t$. That's the impulse (see the section "Looking at the Impact of Impulse" earlier in this chapter), or the force applied to the object multiplied by the time that force was applied. Therefore, you can write this equation as

$$\text{Impulse} = F\Delta t = ma\Delta t = m\Delta v = m(v_f - v_o) = \Delta p$$

Getting rid of everything in the middle finally gives you

$$\text{Impulse} = \Delta p$$

Impulse equals change in momentum. In physics, this process is called the impulse-momentum theorem. The following two sections provide some examples so you can practice this equation.

Shooting pool: Finding impulse and momentum

With the equation Impulse = Δp (see the previous section), you can relate the impulse with which you hit an object to its consequent change in momentum. How about putting yourself to work the next time you hit a pool ball? You line up the shot that the game depends on. You figure that the end of your cue will be in contact with the ball for 5 milliseconds (a millisecond is one one-thousandth of second). How much momentum will the ball need to bounce off the side cushion and end up in the corner pocket?

You measure the ball at 200 grams (or 0.2 kilograms). After testing the side cushion with calipers, spectroscope, and tweezers, killing any chance of finding yourself a date that night, you figure that you need to give the ball a speed of 20.0 meters per second. What average force will you have to apply? To find the average force, you can find the impulse you have to supply. You can relate that impulse to the change in the ball's momentum this way:

$$\text{Impulse} = \Delta p = p_f - p_o$$

So, what's the change in the ball's momentum? The speed you need, 20.0 meters per second, is the magnitude of the pool ball's final velocity. Assuming the pool ball starts at rest, the change in the ball's momentum will be

$$\Delta p = p_f - p_o = m(v_f - v_o)$$

Plugging in the numbers gives you

$$\Delta p = p_f - p_o = m(v_f - v_o) = (0.2)(20 - 0) = 4.0 \text{ kg-m/s}$$

You need a change in momentum of 4.0 kg-m/s, which is also the impulse you need, and because Impulse = FΔt (see the section "Looking at the Impact of Impulse"), this equation becomes

$$F\Delta t = \Delta p = p_f - p_o = m(v_f - v_o) = (0.2)(20 - 0) = 4.0 \text{ kg-m/s}$$

Therefore, the force you need to apply works out to be

$$F = 4.0 \text{ kg-m/s} / \Delta t$$

In this equation, the time your cue ball is in contact with the ball is 5 milliseconds, or 5.0×10^3 seconds, so plugging in that number gives you your desired result:

$$F = 4.0 \text{ kg-m/s} / 5.0 \times 10^3 \text{ s} = 800\text{N}$$

You have to apply about 800N (or about 180 pounds) of force, which seems like a huge amount. However, you apply it over such a short time, 5.0×10^3 seconds, that it seems like much less.

Singing in the rain: An impulsive activity

After a triumphant evening at the pool hall, you decide to leave and discover that it's raining. You grab your umbrella from your car, and the handy rain gauge on the top tells you that 100 grams of water are hitting the umbrella each second at an average speed of 10 meters per second. The question is: If the umbrella has a total mass of 1.0 kg, what force do you need to hold it upright in the rain?

Figuring the force you usually need to hold the weight of the umbrella is no problem — you just figure mass times the acceleration due to gravity, or $(1.0 \text{ kg})(9.8 \text{ m/s}^2) = 9.8\text{N}$. But what about the rain falling on your umbrella? Even if you assume that the water falls off the umbrella immediately, you can't just add the weight of the water, because the rain is falling with a speed of 10 meters per second; in other words, the rain has *momentum*. What can you do? You know that you're facing 100 g of water, or 0.10 kg, falling onto the umbrella each second at a velocity of 10 meters per second downward. When that rain hits your umbrella, the water comes to rest, so the change in momentum per second is

$$\Delta p = m\Delta v$$

Plugging in numbers gives you

$$\Delta p = m\Delta v = (0.1)(10) = 1.0 \text{ kg-m/s}$$

The change in momentum of the rain hitting your umbrella each second is 1.0 kg-m/s. You can relate that to force with the impulse-momentum theorem, which tells you that

Impulse = $F\Delta t = \Delta p$

Dividing both sides by Δt to solve for the force (F) gives you

$F = \Delta p / \Delta t$

You know that $\Delta p = 1.0$ kg-m/s in 1 second, so plugging in Δp and setting Δt to 1 second gives you

$F = \Delta p / \Delta t = 1.0$ kg-m/s / 1.0 second = 1.0 kg-m/s^2 = 1.0 N

In addition to the 9.8N of the umbrella's weight, you also need 1.0N to stand up to the falling rain as it drums on the umbrella, for a total of 10.8N, or about 2.4 pounds of force.

The sticky part of finding force is measuring the small times that are involved in collisions like a cue stick hitting a pool ball. You can remove the time, or Δt, from the process to end up with something a little more useful, which I discuss in the following section.

When Objects Go Bonk: Conserving Momentum

The *principle of conservation of momentum* states that when you have an isolated system with no external forces, the initial total momentum of objects before a collision equals the final total momentum of the objects after the collision ($p_f = p_o$). This people comes out of a bit of algebra and may be the most useful idea I provide in this chapter.

You may have a hard time dealing with the physics of impulses because of the short times and the irregular forces. The absence of complicated external forces is what you need to get a truly useful principle. Troublesome items that are hard to measure — the force and time involved in an impulse — are out of the equation altogether. For example, say that two careless space pilots are zooming toward the scene of an interplanetary crime. In their eagerness to get to the scene first, they collide. During the collision, the average force exerted on first ship by the second ship is F_{12}. By the impulse-momentum theorem (see the section "The Impulse-Momentum Theorem: Relating Impulse and Momentum"), you know the following for the first ship:

$F_{12} \Delta t = \Delta p_1 = m_1\Delta v_1 = m_1(v_{f1} - v_{o1})$

And if the average force exerted on the second ship by the first ship is F_{21}, you also know that

$$F_{21} \Delta t = \Delta p_2 = m_2 \Delta v_2 = m_2(v_{f2} - v_{o2})$$

Now you add these two equations together, which gives you the resulting equation

$$F_{12} \Delta t + F_{21} \Delta t = \Delta p_1 + \Delta p_2 = m_1(v_{f1} - v_{o1}) + m_2(v_{f2} - v_{o2})$$

Rearrange the terms on the right until you get

$$F_{12} \Delta t + F_{21} \Delta t = (m_1 v_{f1} + m_2 v_{f2}) - (m_1 v_{o1} + m_2 v_{o2})$$

This is an interesting result, because $m_1 v_{o1} + m_2 v_{o2}$ is the *initial total momentum* of the two rocket ships, $p_{1o} + p_{2o}$, and $m_1 v_{f1} + m_2 v_{f2}$ is the *final total momentum*, $p_{1f} + p_{2f}$, of the two rocket ships. Therefore, you can write this equation as follows:

$$F_{12} \Delta t + F_{21} \Delta t = (m_1 v_{f1} + m_2 v_{f2}) - (m_1 v_{o1} + m_2 v_{o2}) = (p_{1f} + p_{2f}) - (p_{1o} + p_{2o})$$

If you write the initial total momentum as p_f and the final total momentum as p_o, the equation becomes

$$F_{12} \Delta t + F_{21} \Delta t = (m_1 v_{f1} + m_2 v_{f2}) - (m_1 v_{o1} + m_2 v_{o2}) = p_f - p_o$$

Where do you go from here? You add the two forces together, $F_{12} \Delta t + F_{21}$, to get the sum of the forces involved, ΣF:

$$\Sigma F \Delta t = p_f - p_o$$

If you're working with what's called an *isolated* or *closed system,* you have no external forces to deal with. Such is the case in space. If two rocket ships collide in space, there are no external forces that matter, which means that by Newton's third law (see Chapter 5), $F_{12} = -F_{21}$. In other words, when you have a closed system, you get

$$0 = \Sigma F \Delta t = p_f - p_o$$

This converts to

$$p_f = p_o$$

The equation $p_f = p_o$ says that when you have an isolated system with no external forces, the initial total momentum before a collision equals the final total momentum after a collision, thus giving you the principle of conservation of momentum.

Taking the heat off the final total momentum

When two objects collide and stick together, is the final total momentum really equal to the initial total momentum? Well, not exactly. If one object rams another and the two stick together, a lot of friction could be involved, which means that some kinetic energy goes into heat. When that happens, all bets are off, because the final momentum may not exactly equal the initial momentum. Sometimes in impacts the objects become deformed, as when two cars crash and produce dents. In that case, too, some energy is used in deforming the materials, so the final momentum may not equal the initial momentum. For this reason, friction is usually ignored in introductory physics classes. Otherwise, the problems would get too sticky, because it's very hard to predict how much heat the collision would generate.

Measuring velocity with the conservation of momentum

You can use the principle of conservation of momentum to measure other forces at work, such as velocity. Say, for example, that you're out on a physics expedition, and you happen to pass by a frozen lake where a hockey game is taking place. You measure the speed of one player as 11.0 meters per second just as he collides, rather brutally for a pick-up game, with another player initially at rest. You watch with interest, wondering how fast the resulting mass of hockey players will slide across the ice. After asking a few friends in attendance, you find out that the first player has a mass of 100 kg, and the bulldozed player (who turns out to be his twin) also has a mass of 100 kg. So, what's the final speed of the player tangle?

You're dealing with a closed system (see the previous section), because you neglect the force of friction here, and although the players are exerting a force downward on the ice, the normal force (see Chapter 6) is exerting an equal and opposite force on them, so the vertical force sums to zero.

But what about the resulting horizontal speed along the ice? Due to the principle of conservation of momentum, you know that

$$p_f = p_o$$

Putting this into more helpful terms, you get the following, where you substitute for the initial momentum, which is the momentum of the enforcer, player 1, because the victim isn't moving:

$$p_f = p_o = m_1 v_{o1}$$

The final momentum, p_f, must be equal to the combined mass of the two players multiplied by their velocity, $(m_1 + m_2)v_f$, which gives you

$$p_f = m_1 v_{o1} = (m_1 + m_2)v_f$$

Solving for the final velocity, v_f, gives you

$$v_f = (m_1 v_{o1}) / (m_1 + m_2)$$

Plugging in the numbers gives you

$$v_f = (m_1 v_{o1}) / (m_1 + m_2) = ([100][11]) / (100 + 100) = 1,100 / 200 = 5.5 \text{ meters per second}$$

The speed of the two players together will be half the speed of the original player. That may be what you expected, because you end up with twice the moving mass as before, and because momentum is conserved, you end up with half the speed. Beautiful. You note the results down on your clipboard.

Measuring firing velocity with the conservation of momentum

The principle of conservation of momentum comes in handy when you can't measure velocity with a simple stopwatch. Say, for example, that you accept a consulting job from an ammunition manufacturer that wants to measure the muzzle velocity of its new bullets. No employee has been able to measure the velocity yet, because no stopwatch is fast enough. What will you do? You decide to arrange the setup shown in Figure 9-3, where you fire a bullet of mass m into a hanging wooden block of mass M.

The directors of the ammunition company are perplexed — how can your setup help? Each time you fire a bullet into a hanging wooden block, the bullet kicks the block into the air. So what? You decide they need a lesson on the principle of conservation of momentum. The original momentum, you explain, is the momentum of the bullet:

$$p_o = mv_o$$

Because the bullet sticks into the wood block, the final momentum is the product of the total mass, $m + M$, and the final velocity of the bullet/wood block combination:

$$p_f = (m + M)v_f$$

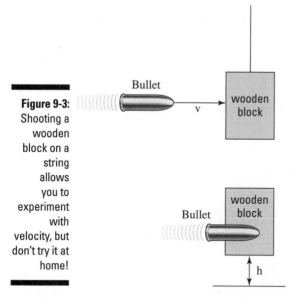

Figure 9-3:
Shooting a
wooden
block on a
string
allows
you to
experiment
with
velocity, but
don't try it at
home!

Because of the principle of conservation of momentum and neglecting friction and other nonconservative forces, you can say that

$$p_f = p_o$$

So,

$$v_f = mv_o / (m + M)$$

The directors start to get dizzy, so you explain how the kinetic energy of the block when it's struck goes into its final potential energy when it rises to height *h* (see Chapter 8), so

$$\tfrac{1}{2}(m + M)v_f^2 = (m + M)gh$$

Putting all the numbers together gives you

$$\tfrac{1}{2}(m + M)v_f^2 = \tfrac{1}{2}(m + M)(m^2 v_o^2 / [m + M]^2) = (m + M)gh$$

With a flourish, you add that solving for the initial velocity, v_o, gives you

$$v_o = \sqrt{\frac{2(m + M)^2 gh}{m^2}}$$

You measure that the bullet has a mass of 50 g, the wooden block has a mass of 10 kg, and that upon impact, the block rises 50.0 cm into the air. Plugging in those values gives you your result:

$$v_0 = \sqrt{\frac{2(m+M)^2 gh}{m^2}} = 629 \text{ meters per second}$$

The initial velocity is 629 meters per second, which converts to about 2,060 feet per second. "Brilliant!" the directors cry as they hand you a big check.

When Worlds (or Cars) Collide: Elastic and Inelastic Collisions

Examining collisions in physics can be pretty entertaining, especially because the principle of conservation of momentum makes your job so easy (see the section "When Objects Go Bonk: Conserving Momentum" earlier in this chapter to find out how). But there's often more to the story when you're dealing with collisions than impulse and momentum. Sometimes, kinetic energy is also conserved, which gives you the extra edge you need to figure out what happens in all kinds of collisions, even across two dimensions.

You can run into all kinds of situations in physics problems where collisions become important. Two cars collide, for example, and you need to find the final velocity of the two when they stick together. You may even run into a case where two railway cars going at different velocities collide and couple together, and you need to determine the final velocity of the two cars.

But what if you have a more general case where the two objects don't stick together? Say, for example, you have two pool balls that hit each other at different speeds and at different angles and bounce off with different speeds and different angles. How the heck do you handle that situation? You have a way to handle these collisions, but you need more than just the principle of conservation of momentum gives you.

When objects bounce: Elastic collisions

When bodies collide in the real world, you can observe energy losses due to heat and deformation. If these losses are much smaller than the other energies involved, such as when two pool balls collide and go their separate ways, kinetic energy may be conserved in the collision. Physics has a special name for collisions where kinetic energy is conserved: *elastic collisions*. In an elastic collision, the total kinetic energy in a closed system (where the net forces add up to zero) is the same before the collision as after the collision.

When objects don't bounce: Inelastic collisions

If you can observe appreciable energy losses due to nonconservative forces (such as friction) during a collision, kinetic energy isn't conserved. In this case, friction, deformation, or some other process transforms the kinetic energy, and it's lost. The name physics gives to a situation where kinetic energy is lost after a collision is an *inelastic collision*. The total kinetic energy in a closed system isn't the same before the collision as after the collision. You see inelastic collisions when objects stick together after colliding, such as when two cars crash and weld themselves into one.

Objects don't need to stick together in an inelastic collision; all that has to happen is the loss of some kinetic energy. For example, if you smash into a car and deform it, the collision is inelastic, even if you can drive away after the accident.

Colliding along a line

When a collision is elastic, kinetic energy is conserved. The most basic way to look at elastic collisions is to examine how the collisions work along a straight line. If you smash your bumper car into a friend's bumper car along a straight line, you bounce off and kinetic energy is conserved along the line. But the behavior of the cars depends on the mass of the objects involved in the elastic collision.

Bumping into a heavier mass

You take your family to the Physics Amusement Park for a day of fun and calculation, and you decide to ride the bumper cars. You wave to your family as you speed your 300-kg car up to 10 meters per second. Suddenly, BONK! What happened? The person in front of you, driving a 400-kg car came to a complete stop, and you rear-end him elastically; now you're traveling backward and he's traveling forward. "Interesting," you think. "I wonder if I can solve for the final velocities of both bumper cars."

You know that the momentum was conserved, and you know that the car in front of you was stopped when you hit it, so if your car is car 1 and the other is car 2, you get the following

$$m_1 v_{f1} + m_2 v_{f2} = m_1 v_{o1}$$

However, this doesn't tell you what v_{f1} and v_{f2} are, because there are two unknowns and only one equation here. You can't solve for v_{f1} or v_{f2} exactly in this case, even if you know the masses and v_{o1}. You need some other equations relating these quantities. How about using the conservation of kinetic energy? The collision was elastic, so kinetic energy was conserved, which means that

$$\tfrac{1}{2}m_1v_{f1}^2 + \tfrac{1}{2}m_2v_{f2}^2 = \tfrac{1}{2}m_1v_{o1}^2$$

Now you have two equations and two unknowns, v_{f1} and v_{f2}, which means you can solve for the unknowns in terms of the masses and v_{o1}. You have to dig through a lot of algebra here because the second equation has many squared velocities, but when the dust settles, you get

$$v_{f1} = [(m_1 - m_2)v_{o1}] / (m_1 + m_2)$$

and

$$v_{f2} = 2m_1v_{o1} / (m_1 + m_2)$$

Now you have v_{f1} and v_{f2} in terms of the masses and v_{o1}. Plugging in the numbers gives you

$$v_{f1} = [(m_1 - m_2)v_{o1}] / (m_1 + m_2) = [(300 - 400)(10)] / (300 + 400) = -1.43 \text{ m/s}$$

and

$$v_{f2} = 2m_1v_{o1} / (m_1 + m_2) = [2(300)(10)] / (300 + 400) = 8.57 \text{ m/s}$$

The two speeds tell the whole story. You started off at 10.0 meters per second in a bumper car of 300 kg, and you hit a stationary bumper car of 400 kg in front of you. Assuming the collision took place directly and the second bumper car took off in the same direction you were going before the collision, you rebounded at –1.43 meters per second — backward, because this quantity is negative and the bumper car in front of you had more mass — and the bumper car in front of you took off at a speed of 8.57 meters per second.

Bumping into a lighter mass

After having a bad experience in a previous trip to the bumper car pit — where your light bumper car rear-ended a heavy bumper car (see the previous section for the calculation) — you decide to go back and pick on some poor light cars in a monster bumper car. What happens if your bumper car has a mass of 400 kg, and you rear-end a stationary 300-kg car? In this case, you use the equation for conservation of kinetic energy, the same formula you use in the previous section. Your final velocity comes out to

$$v_{f1} = [(m_1 - m_2)v_{o1}] / (m_1 + m_2) = [(400 - 300)(10)] / (300 + 400) = 1.43 \text{ m/s}$$

The little car's final velocity comes out to

$$v_{f2} = 2m_1v_{o1} / (m_1 + m_2) = [2(400)(10)] / (300 + 400) = 11.4 \text{ m/s}$$

In this case, you don't bounce backward. The lighter, stationary car takes off after you hit it, but not all your forward momentum is transferred to the other car. Is momentum still conserved? Take a look at the following:

$$p_o = m_1 v_{o1}$$

and

$$p_f = m_1 v_{f1} + m_2 v_{f2}$$

Putting in the numbers, you get

$$p_o = m_1 v_{o1} = (400)(10) = 4,000 \text{ kg-m/s}$$

and

$$p_f = m_1 v_{f1} + m_2 v_{f2} = (400)(1.43) + (300)(11.4) = 4,000 \text{ kg-m/s}$$

Momentum is conserved in this collision, just as it is for your collision with a heavier car.

Colliding in two dimensions

Collisions don't always occur along a straight line. For example, balls on a pool table can go in two dimensions, both X and Y, as they zoom around. Collisions along two dimensions introduce variables such as angle and direction. Say, for example, your physics travels take you to the golf course, where two players are lining up for their final putts of the day. The players are tied, so these putts are the deciding shots. Unfortunately, the player closer to the hole breaks etiquette, and they both putt at the same time. Their 45-g golf balls collide! You can see what happens in Figure 9-4.

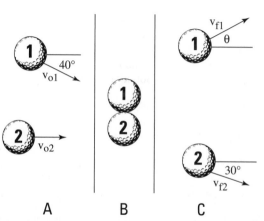

Figure 9-4:
Before, during, and after a collision between two balls moving in two dimensions.

A   B   C

You quickly stoop down to measure all the angles and velocities involved in the collision. You measure the speeds: $v_{o1} = 1.0$ meter per second, $v_{o2} = 2.0$ meters per second, and $v_{f2} = 1.2$ meters per second. You also get most of the angles, as shown in Figure 9-4. However, you can't get the final angle and speed of golf ball 1.

Because the golf balls create an elastic collision, both momentum and kinetic energy are conserved. In particular, momentum is conserved in both the X-axis and Y-axis directions, and total kinetic energy is conserved as well. But using all those equations in two dimensions can be a nightmare, so physics problems very rarely ask you to do that kind of calculation. In this case, all you want is the final velocity — that is, the speed and direction — of golf ball 1, as shown in Figure 9-4. To solve this problem, all you need is the conservation of momentum in two dimensions. Momentum is conserved in both the X and Y directions, which means that

$$p_{fx} = p_{ox}$$

and

$$p_{fy} = p_{oy}$$

In other words, the final momentum in the X direction is the same as the original momentum in the X direction, and the final momentum in the Y direction is the same as the original momentum in the Y direction. Here's what the original momentum in the X direction looks like:

$$p_{fx} = p_{ox} = m_1 v_{o1} \cos 40° + m_2 v_{o1}$$

Setting that equal to the final momentum in the X direction gives you

$$p_{fx} = p_{ox} = m_1 v_{o1} \cos 40° + m_2 v_{o2} = m_1 v_{f1x} + m_2 v_{f2} \cos 30°$$

This tells you that

$$m_1 v_{f1x} = m_1 v_{o1} \cos 40° + m_2 v_{o2} - m_2 v_{f2} \cos 30°$$

Dividing by m_1 gives you

$$v_{f1x} = v_{o1} \cos 40° + (m_2 v_{o2} - m_2 v_{f2} \cos 30°) / m_1$$

Because $m_1 = m_2$, this breaks down even more:

$$v_{f1x} = v_{o1} \cos 40° + v_{o2} - v_{f2} \cos 30°$$

Plugging in the numbers gives you

$$v_{f1x} = v_{o1} \cos 40° + v_{o2} - v_{f2} \cos 30° = (1.0)(.766) + 2.0 - [(1.2)(.866)] = 1.71 \text{ m/s}$$

The final velocity of golf ball 1 in the X direction is 1.71 meters per second.

Chapter 10

Winding Up with Angular Kinetics

. . . .

In This Chapter

▶ Changing gears from linear motion to rotational motion

▶ Calculating tangential speed and acceleration

▶ Figuring angular acceleration and velocity

▶ Identifying the torque involved in rotational motion

▶ Maintaining rotational equilibrium

. . . .

*T*his chapter is the first of two (Chapter 11 is the other) on handling objects that rotate, from space stations to marbles. Rotation is what makes the world go round — literally — and if you know how to handle linear motion and Newton's laws (see the first two parts of the book if you don't), the rotational equivalents I present in this chapter and in Chapter 11 are pieces of cake. And if you don't have a grasp on linear motion, no worries. You can get a firm grip on the basics of rotation here and go back for the linear stuff. You see all kinds of rotational ideas in this chapter: angular acceleration, tangential speed and acceleration, torque, and more. But enough spinning the wheels. Read on!

Going from Linear to Rotational Motion

You need to change equations when you go from linear motion to rotational motion, particularly when angles get involved. Chapter 7 shows you the rotational equivalents (or *analogs*) for each of these linear equations:

▶ $v = \Delta s / \Delta t$, where v is velocity, Δs is the change in displacement, and Δt is the change in time

▶ $a = \Delta v / \Delta t$, where a is acceleration and Δv is the change in velocity

▶ $s = v_0(t_f - t_o) + \frac{1}{2}a(t_f - t_o)^2$, where s is displacement, v_0 is the original velocity, t_f is the final time, and t_o is the original time

▶ $v_f^2 - v_o^2 = 2as$, where v_f is the final velocity

Here's how you convert these equations in terms of angular displacement, θ (measured in radians — 2π radians in a circle), angular speed, ω, and angular acceleration, α:

$$\omega = \Delta\theta \, / \, \Delta t$$

$$\alpha = \Delta\omega \, / \, \Delta t$$

$$\theta = \omega_0(t_f - t_o) + \tfrac{1}{2}\alpha \, (t_f - t_o)^2$$

$$\omega_f^2 - \omega_o^2 = 2\alpha\theta$$

Understanding Tangential Motion

Tangential motion is motion that's perpendicular to radial motion, or motion along a radius. You can tie angular quantities like angular displacement, θ, angular speed, ω, and angular acceleration, α, to their associated tangential quantities — all you have to do is multiply by the radius:

$$s = r\theta$$

$$v = r\omega$$

$$a = r\alpha$$

Say you're riding a motorcycle, for example, and the wheels' final angular speed is $\omega_f = 21.5\pi$ radians per second (see the previous section for the full calculation). What does this mean in terms of your motorcycle's speed? To determine your motorcycle's speed, you need to relate angular speed, ω, to linear speed, v. The following sections explain how you can make such relations.

Finding tangential speed

Linear speed has a special name when you begin to deal with rotational motion — it's called the tangential speed. *Tangential speed* is the speed of a point at a given radius r. The vector v shown in Figure 10-1 is a tangential vector (meaning it has a magnitude and a direction; see Chapter 4), perpendicular to the radius.

Given an angular speed ω, the tangential speed at any radius is $r\omega$. This makes sense, because given a rotating wheel, you'd expect a point at radius r to be going faster than a point closer to the hub of the wheel.

Take a look at Figure 10-1, which shows a ball tied to a string. In this case, you know that the ball is whipping around with angular speed ω.

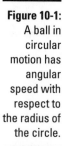

Figure 10-1:
A ball in
circular
motion has
angular
speed with
respect to
the radius of
the circle.

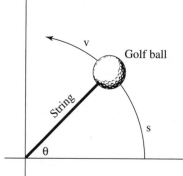

You can easily find the magnitude of the ball's velocity, v, if you measure the angles in radians. A circle has 2π radians; the complete distance around a circle — its circumference — is $2\pi r$, where r is the circle's radius. And if you go only halfway around, you cover a distance of πr, or π radians. In general, therefore, you can connect an angle measured in radians with the distance you cover along the circle, s, like this:

$$s = r\theta$$

where r is the radius of the circle. Now, you can say that

$$v = s / t$$

where v is velocity, s is the displacement, and t is time. You can substitute for s to get

$$v = s / t = r\theta / t$$

$\omega = \theta / t$, which means

$$v = s / t = r\theta / t = r\omega$$

In other words,

$$v = r\omega$$

Now you can find the magnitude of the velocity. The wheels of the motorcycle are turning with an angular speed of 21.5π radians per second. If you can find the tangential speed of any point on the outside edges of the wheels, you can find the motorcycle's speed. Say, for example, that the radius of one of your motorcycle's wheels is 40 cm. You know that

$$v = r\omega$$

Just plug in the numbers!

$$v = r\omega = (0.40)(21.5\pi) = 27.0 \text{ m/s}$$

Converting 27.0 meters per second to miles per hour gives you about 60 mph.

Finding tangential acceleration

Tangential acceleration is a measure of how the speed of a point at a certain radius changes with time. This type of acceleration resembles linear acceleration (see Chapter 3), with the exception that tangential acceleration is all about circular motion. For example, when you start a lawn mower, a point on the tip of one of its blades starts at a tangential speed of zero and ends up with a pretty fast tangential speed. So, how do you determine the point's tangential acceleration? How can you relate the following equation from Chapter 3, which finds linear acceleration (where Δv is the change in velocity and Δt is the change in time)

$$a = \Delta v / \Delta t$$

to angular quantities like angular speed? You find in the previous section that tangential speed, v, equals $r\omega$, so you can plug this information in:

$$a = \Delta v / \Delta t = \Delta(r\omega) / \Delta t$$

Because the radius is constant here, the equation becomes

$$a = \Delta v / \Delta t = \Delta(r\omega) / \Delta t = r\Delta\omega / \Delta t$$

However, $\Delta\omega / \Delta t = \alpha$, the angular acceleration, so the equation becomes

$$a = \Delta v / \Delta t = r\Delta\omega / \Delta t = r\alpha$$

In other words,

$$a = r\alpha$$

Translated into laymen's terms, this says tangential acceleration equals angular acceleration multiplied by the radius.

Finding centripetal acceleration

Another kind of acceleration turns up in an object's circular motion — *centripetal acceleration,* or the acceleration an object needs to keep going in a circle. Can you connect angular quantities, like angular speed, to centripetal acceleration? You sure can. Centripetal acceleration is given by the following equation (for more on the equation, see Chapter 7):

$$a_c = v^2 / r$$

where v^2 is velocity squared and r is the radius. This is easy enough to tie to angular speed because $v = r\omega$ (see the section "Finding tangential speed"), which gives you

$$a_c = v^2 / r = (r\omega)^2 / r$$

This equation breaks down to

$$a_c = r\omega^2$$

Nothing to it. The equation for centripetal acceleration means that you can find the centripetal acceleration needed to keep an object moving in a circle given the circle's radius and the object's angular speed. For example, say that you want to calculate the centripetal acceleration of the moon around the Earth. First, use the old equation:

$$a_c = v^2 / r$$

First you have to calculate the tangential velocity, or angular speed, of the moon in its orbit. Using the new version of the equation, $a_c = r\omega^2$, is easier because the moon orbits the Earth in about 28 days, so you can easily calculate the moon's angular speed.

Because the moon makes a complete orbit around the earth in about 28 days, it travels 2π radians around the Earth in that period, so its angular speed is

$$\omega = \Delta\theta / \Delta t = 2\pi / 28 \text{ days}$$

Converting 28 days to seconds gives you (28 days)(24 hours per day)(60 minutes per hour)(60 seconds per minute) = 2.42×10^6 seconds, so you get the following angular speed:

$$\omega = \Delta\theta / \Delta t = 2\pi / 28 \text{ days} = 2\pi / 2.42 \times 10^6 = 2.60 \times 10^{-6} \text{ radians per second}$$

You now have the moon's angular speed, 2.60×10^{-6} radians per second. The average radius of the moon's orbit is 3.85×10^8 m, so its centripetal acceleration is

$$a_c = r\omega^2 = (3.85 \times 10^8)(2.60 \times 10^{-6})^2 = 2.60 \times 10^{-3} \text{ radians per second}^2$$

Just for kicks, you can also find the force needed to keep the moon going around in its orbit. Force equals mass times acceleration (see Chapter 5), so you multiply by the mass of the moon, 7.35×10^{22} kg:

$$F_c = ma_c = (7.35 \times 10^{22})(2.60 \times 10^{3}) = 1.91 \times 10^{20} \text{N}$$

The force in Newtons you find, 1.91×10^{20}N, converts to about 4.30×10^{19} pounds of force needed to keep the moon going around in its orbit.

Applying Vectors to Rotation

The previous sections in this chapter examine angular speed and angular acceleration as if they're scalars — as if speed and acceleration have only a magnitude and not a direction. However, these concepts are really vectors, which means they have a magnitude and a direction (see Chapter 4 for more on scalars and vectors). When you make the switch from linear motion to circular motion, you make the switch from angular speed to angular velocity — and from talking about only magnitudes to magnitudes *and* direction. You can see the relation between the two in the following sections.

Angular velocity and angular acceleration are vectors that point at right angles to the direction of rotation.

Calculating angular velocity

When a wheel is spinning, it has an angular speed, but it can have an angular acceleration as well. Say, for example, that the wheel has a constant angular speed, ω — which direction does its angular velocity, ω, point? It can't point along the rim of the wheel, as tangential velocity does, because its direction would change every second. In fact, the only real choice for its direction is perpendicular to the wheel.

This always takes people by surprise: angular velocity, ω, points along the axle of a wheel (for an example, see Figure 10-2). Because the angular velocity vector points the way it does, it has no component along the wheel. The wheel is spinning, so the velocity at any point on the wheel is constantly changing direction — except for the very center point of the wheel. The angular velocity vector's base sits at the very center point of the wheel. Its head points up or down, away from the wheel.

You can use the *right hand rule* to determine a vector's direction. To apply this rule to the wheel in Figure 10-2, wrap your right hand (left-handed people often think that right-handed chauvinists invented this rule, and maybe that's true) around the wheel so that your fingers point in the direction of the tangential motion at any point — that is, the fingers on your right hand go in the same direction as the wheel's rotation. When you wrap your right hand around the wheel, your thumb will point in the direction of the angular velocity vector, ω.

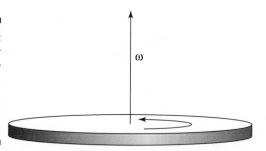

Figure 10-2:
Angular velocity points in a perpendicular direction to the circle.

Now you can master the angular velocity vector. You know that its magnitude is ω, the angular speed of an object in rotational motion. And now you can find the direction of that vector by using the right hand rule. The fact that the angular velocity is perpendicular to the plane of rotational motion (the flat side of the wheel) takes some getting used to, but as you've seen, you can't plant a vector on a spinning wheel that has constant angular velocity so that the vector has a constant direction, except at the very center of the wheel. And from there, you have no way to go except up (or down, in the case of negative angular velocity).

Figuring angular acceleration

If the angular velocity vector points out of the plane of rotation (see the previous section), what happens when the angular velocity changes — when the wheel speeds up or slows down? A change in velocity signifies the presence of angular acceleration. Like angular velocity, ω, angular acceleration, α, is a vector, meaning it has a magnitude and a direction. Angular acceleration is the rate of change of angular speed:

$$\alpha = \Delta\omega \,/\, \Delta t$$

For example, take a look at Figure 10-3, which shows what happens when angular acceleration affects angular velocity.

In this case, α points in the same direction as ω in diagram A. If the angular acceleration vector, α, points along the angular velocity, ω, as time goes on, the magnitude of ω will increase, as shown in Figure 10-3, diagram B.

Figure 10-3:
The angular acceleration vector indicates how angular velocity will change in time.

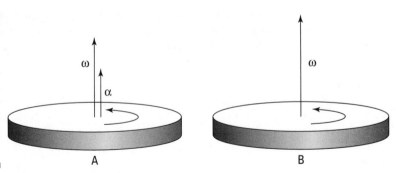

A   B

You've calculated that the angular acceleration vector just indicates how the angular velocity will change in time, which mirrors the relationship between linear acceleration and linear velocity. However, you should note that the angular acceleration doesn't have to be in the same direction as the angular velocity vector at all, as shown in Figure 10-4, diagram A. If the angular acceleration moves in the opposite direction of the angular velocity, it's called *negative angular acceleration*.

Figure 10-4:
Spinning in the opposite direction of angular velocity with negative angular acceleration.

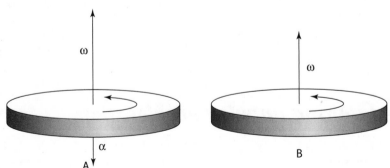

A   B

As you may expect in this case, the angular acceleration will reduce the angular velocity as time goes on, which you can see in Figure 10-4, diagram B. Say, for example, that you grab the spinning wheel's axle in Figure 10-3 and tip the wheel. The angular acceleration is at right angles to the angular velocity, which changes the direction of the angular velocity.

Twisting and Shouting: Torque

You not only have to look at how forces work when you apply them to objects, but also *where* you apply the forces. Enter torque. *Torque* is a measure of the tendency of a force to cause rotation. In physics terms, the torque exerted on an object depends on where you exert the force. You go from the strictly linear idea of force as something that acts in a straight line, such as pushing a refrigerator up a ramp, to its angular counterpart, torque.

Torque brings forces into the rotational world. Most objects aren't just point or rigid masses, so if you push them, they not only move but also turn. For example, if you apply a force tangentially to a merry-go-round, you don't move the merry-go-round away from its current location — you cause it to start spinning. And spinning is the rotational kinematic you focus on this chapter and in Chapter 11.

Take a look at Figure 10-5, which shows a seesaw with a mass *m* on it. If you want to balance the seesaw, you can't have a larger mass, M, placed on a similar spot on the other side of the seesaw. Where you put the larger mass M determines what results you get. As you can see in diagram A of Figure 10-5, if you put the mass M on the *pivot point* — also called the *fulcrum* — of the seesaw, you don't create a balance. The larger mass exerts a force on the seesaw, but the force doesn't balance it.

As you can see in diagram B of Figure 10-5, as the distance you put the mass M away from the fulcrum increases, the balance improves. In fact, if M = 2m, you need to put the mass M exactly half as far from the fulcrum as the mass *m* is.

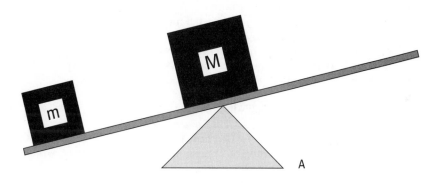

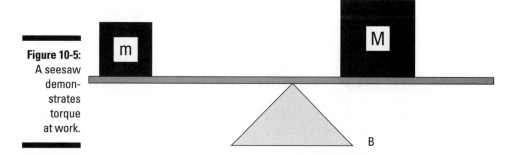

Figure 10-5:
A seesaw demon-
strates
torque
at work.

Mapping out the torque equation

How much torque you exert on an object depends on the point where you apply the force. The force you exert, F, is important, but you can't discount the *lever arm* — also called the *moment arm* — which is the distance from the pivot point at which you exert your force. Assume that you're trying to open a door, as in the various scenarios in Figure 10-6. You know that if you push on the hinge, as in diagram A, the door won't open; if you push the middle of the door, as in diagram B, the door will open slowly; and if you push the edge of the door, as in diagram C, the door will open faster.

In Figure 10-6, the lever arm, l, is distance r from the hinge at which you exert your force. The torque is the product of the force multiplied by the lever arm. It has a special symbol, the Greek letter τ (tau):

$$\tau = Fl$$

The units of torque are force multiplied by distance, which is Newtons-meters in the MKS system, dynes-centimeters in the CGS system, and feet-pounds in the foot-pound-second system (see Chapter 2 for more on these measurement systems).

So, for example, the lever arm in Figure 10-6 is distance r, so $\tau = Fr$. If you push with a force of 200N, and $r = 0.5$ meters, what's the torque you see in the figure? In diagram A, you push on the hinge, so your distance from the pivot point is zero, which means the lever arm is zero. Therefore, the torque is zero. In diagram B, you exert the 200N of force at a distance of 0.5 meters perpendicular to the hinge, so

$$\tau = Fl = 200(0.5) = 100\text{N-m}$$

The torque here is 100 N-m. But now take a look at diagram C. You push with 200N of force at a distance of 2r perpendicular to the hinge, which makes the lever arm 2r = 1.0 meter, so you get this torque:

$$\tau = Fl = 200(1.0) = 200\text{N-m}$$

Now you have 200 N-m of torque, because you push at a point twice as far away from the pivot point. In other words, you double your torque. But what would happen if, say, the door were partially open when you exerted your force? Well, you would calculate the torque easily, if you have lever-arm mastery.

Understanding lever arms

If you push a partially open door in the same direction as you push a closed door, you create a different torque because of the non-right angle between your force and the door.

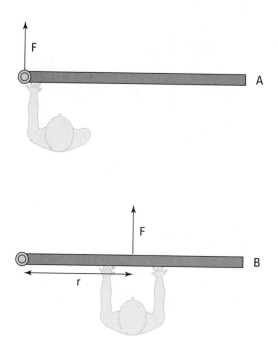

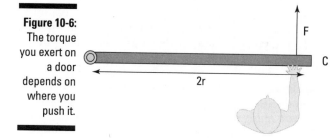

Figure 10-6:
The torque
you exert on
a door
depends on
where you
push it.

Take a look at Figure 10-7, diagram A, to see a person obstinately trying to open a door by pushing along the door toward the hinge. You know this method won't produce any turning motion, because the person's force has no lever arm to produce the needed turning force. "Leave me alone," the person says; some people just don't appreciate physics. In this case, the lever arm was zero, so it's clear that even if you apply a force at a given distance away from a pivot point, you don't always produce a torque. The direction you apply the force also counts, as you know from your door-opening expertise.

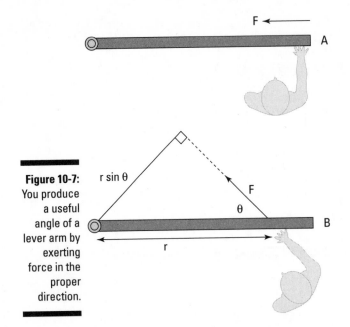

Figure 10-7:
You produce
a useful
angle of a
lever arm by
exerting
force in the
proper
direction.

Figuring out the torque generated

Generating torque is how you open doors, whether you have to quickly pop a car door or slowly pry open a bank-vault door. But how do you find out how much torque you generate? First, you calculate the lever arm, and then you multiply that lever arm by the force to get the torque.

Take a look at diagram B in Figure 10-7. You apply a force to the door at some angle, θ. The force may open the door, but it isn't a sure thing, because as you can tell from the figure, you apply less of a turning force here. What you need to do is find the lever arm first. As you can see in Figure 10-7, you apply the force at a distance r from the hinge. If you apply that force perpendicularly to the door, the lever arm would be r, and you'd get

$$\tau = Fr$$

However, that's not the case here, because the force isn't perpendicular to the door.

The lever arm is the effective distance from the pivot point at which the force acts perpendicularly.

To see how this works, take a look at diagram B in Figure 10-7, where you can draw a lever arm from the pivot point so that the force is perpendicular to the lever arm. To do this, extend the force vector until you can draw a line from the pivot point that's perpendicular to the force vector. You create a new triangle. The lever arm and the force are at right angles with respect to each other, so you create a right triangle. The angle between the force and the door is θ, and the distance from the hinge at which you apply the force is r (the hypotenuse of the right triangle), so the lever arm becomes

$$l = r \sin \theta$$

When θ goes to zero, so does the lever arm, so there's no torque (see diagram A in Figure 10-7). You know that

$$\tau = Fl$$

so you can now find

$$\tau = Fr \sin \theta$$

where θ is the angle between the force and the door.

This is a general equation; if you apply a force F at a distance r from a pivot point, where the angle between that displacement and F is θ, the torque you produce will be $\tau = Fr \sin \theta$. So, for example, if $\theta = 45°$, F = 200N, and $r = 1.0$ meter, you get

$$\tau = Fr \sin \theta = (200)(1.0)(0.707) = 141 \text{N-m}$$

This number is less than you'd expect if you just push perpendicularly to the door (which would be 200 N-m).

Recognizing that torque is a vector

Looking at angles between lever arms and force vectors (see the previous two sections) may tip you off that torque is a vector, too. And it is. In physics, torque is a positive vector if it tends to create a counterclockwise turning motion (toward increasingly positive angles) and negative if it tends to create a clockwise turning motion (toward increasingly negative angles).

For example, take a look at Figure 10-8, where a force F applied at lever arm l is producing a torque τ. Because the turning motion produced is toward larger positive angles, τ is positive.

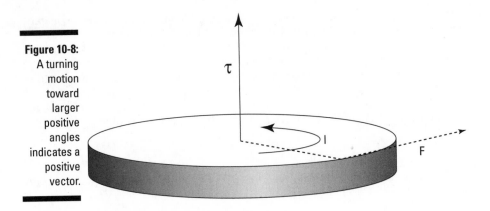

Figure 10-8:
A turning
motion
toward
larger
positive
angles
indicates a
positive
vector.

No Wobbling Allowed: Rotational Equilibrium

You may know equilibrium as a state of balance, but what's equilibrium in physics terms? When you say an object has *equilibrium,* you mean that the motion of the object isn't changing; in other words, the object has no acceleration (it can have motion, however, as in constant velocity and/or constant angular velocity). As far as linear motion goes, the sum of all forces acting on the object must be zero:

$$\Sigma F = 0$$

In other words, the net force acting on the object is zero.

Equilibrium also occurs in rotational motion in the form of rotational equilibrium. When an object is in *rotational equilibrium,* it has no angular acceleration — the object may be rotating, but it isn't speeding up or slowing down, which means its angular velocity is constant. When an object has rotational equilibrium, there's no net turning force on the object, which means that the net torque on the object must be zero:

$$\Sigma \tau = 0$$

This equation represents the rotational equivalent of linear equilibrium. Rotational equilibrium is a useful idea, because given a set of torques operating on an object, you can determine what torque is necessary to stop the object from rotating.

Hanging a flag: A rotational equilibrium problem

The manager at the hardware store you work at asks you to help hang a flag over the top of the store. The store is extra-proud of the flag, because it's an extra-big one (to check it out, see Figure 10-9). The problem is that the bolt holding the flagpole in place seems to break all the time, and both the flag and pole go hurtling over the edge of the building, which doesn't help the store's image.

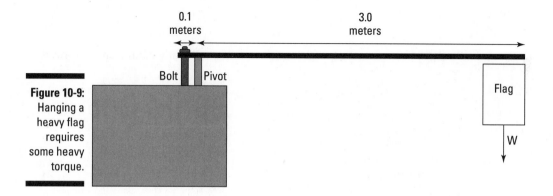

Figure 10-9: Hanging a heavy flag requires some heavy torque.

In order to find out how much force the bolt needs to provide, you start taking measurements and note that the flag has a mass of 50 kg — much more than the mass of the pole — and that the manager had previously hung it 3.0 meters from the pivot point. The bolt is 10 cm from the pivot point. To get rotational equilibrium, you need to have zero net torque:

$$\Sigma\tau = 0$$

In other words, if the torque due to the flag is τ_1 and the torque due to the bolt is τ_2,

$$\Sigma\tau = 0 = \tau_1 + \tau_2$$

What are the torques involved here? You know that the flag's weight provides a torque around the pivot point, τ_1, so

$$\tau_1 = mgl_1$$

where m is the mass of the pole, g is the acceleration due to gravity, and l_l is the lever arm for the flag. Plugging in the numbers gives you

$$\tau_1 = mgl_1 = (50)(-9.8)(3.0) = -1,470\text{N-m}$$

Note that this is a negative torque because the acceleration due to gravity is pointing down. What about the torque due to the bolt, τ_2? As with any torque, you can write τ_2 as

$$\tau_2 = F_2l_2$$

Plugging in as many numbers as you know gives you

$$\tau_2 = F_2l_2 = F_2(0.10)$$

Because you want rotational equilibrium, the following condition must hold:

$$\Sigma\tau = 0 = \tau_1 + \tau_2$$

In other words, the torques must balance out, so

$$\tau_2 = -\tau_1$$

which means that

$$\tau_2 = -\tau_1 = 1,470\text{N-m}$$

Now you can finally get your answer:

$$\tau_2 = F_2l_2 = F_2(0.10) = 1,470\text{N-m}$$

Putting F_2 on one side and solving the equation gives you

$$F_2(0.10) / 0.10 = 1,470\text{N-m} / 0.10 = 14,700\text{N}$$

The bolt needs to provide at least 14,700N of force, or about 330 pounds.

Ladder safety: Introducing friction into rotational equilibrium

The hardware store owner from the previous section has come to you again for help with another problem. A clerk has climbed near the top of a ladder to hang a sign for the company's upcoming sale. The owner doesn't want the ladder to slip — lawsuits, he explains — so he asks you if the ladder is going to fall.

The situation appears in Figure 10-10. The question is, will the force of friction keep the ladder from moving if θ is 45° and the static coefficient of friction (see Chapter 6) with the floor is 0.7?

You have to work with net forces to determine the overall torque. You write down what you know (you can assume that the weight of the ladder is concentrated at its middle):

- F_W = Force exerted by the wall on the ladder
- W_C = Weight of the clerk = 450N
- W_L = Weight of the ladder = 200N
- F_F = Force of friction holding the ladder in place
- F_N = Normal force (see Chapter 6)

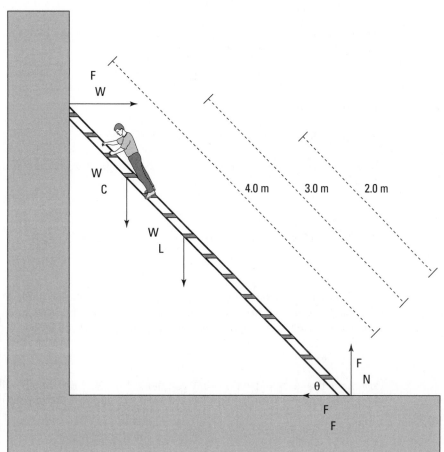

Figure 10-10:
Keeping a
ladder
upright
requires
friction and
rotational
equilibrium.

You need to determine the needed force of friction here, and you want the ladder to be in both linear and rotational equilibrium. Linear equilibrium tells you that the force exerted by the wall on the ladder, F_W, must be the same as the force of friction in magnitude but opposite in direction, because those are the only two horizontal forces. Therefore, if you can find F_W, you know what the force of friction, F_F, needs to be, which is your goal here.

You know that the ladder is in rotational equilibrium, which means that

$$\Sigma \tau = 0$$

To find F_W, take a look at the torques around the bottom of the ladder, using that point as the pivot point. All the torques around the pivot point have to add up to zero. The torque due to the force from the wall against the ladder (see the section "Figuring out the torque generated" for more on angles) is

$$-F_W \, (4.0m) \sin 45° = -2.83 \, F_W$$

Note that F_W is negative, because it tends to produce a clockwise motion. The torque due to the clerk's weight is

$$W_c \, (3.0m) \cos 45° = (450N)(3.0m)(0.707) = 954 \text{N-m}$$

The torque due to the ladder's weight is

$$W_L \, (4.0m) \cos 45° = (200N)(2.0m)(0.707) = 283 \text{N-m}$$

Both of these torques are positive; because $\Sigma \tau = 0$, you get the following result when you add all the torques together:

$$\Sigma \tau = 954 \text{ N-m} + 283 \text{ N-m} - 2.83 \, F_W = 0$$

You can add 2.83 F_W to each side and then divide by 2.83 to get

$$F_W = (954 \text{ N-m} + 283 \text{ N-m}) / 2.83 = 437N$$

The force the wall exerts on the ladder is 437N, which is also equal to the frictional force of the bottom of the ladder on the floor, because F_W and the frictional force are the only two horizontal forces in the whole system. Therefore,

$$F_F = 437N$$

You have the force of friction that you need. But what do you actually have? The basic equation for friction (as outlined in Chapter 6) tells you that

$$F_{F \, actual} = \mu_s \, F_N$$

where μ_s is the coefficient of static friction and F_N is the normal force of the floor pushing up on the ladder, which must balance all the downward-pointing forces in this problem because of linear equilibrium. This means that

$$F_N = W_C + W_L = 450N + 200N = 650N$$

Plugging this into the equation for $F_{F\,actual}$ and using the value of μ_s, 0.7, gets you the following:

$$F_{F\,actual} = \mu_s\, F_N = (0.7)(650) = 455N$$

You need 437N of force, and you actually have 455N. Good news — the ladder isn't going to slip.

Chapter 11

Round and Round with Rotational Dynamics

This chapter is all about applying forces and seeing what happens in the rotational world. You find out what Newton's second law (force equals mass times acceleration; see Chapter 5) becomes for rotational motion, you see how inertia comes into play in rotational motion, and you get the story on rotational kinetic energy, rotational work, and angular momentum. All that rolls comes up in this chapter, and you get the goods on it.

Rolling Up Newton's Second Law into Angular Motion

Newton's second law, force equals mass times acceleration (F = *ma;* see Chapter 5), is a physics favorite in the linear world because it ties together the vectors force and acceleration (see Chapter 4 for more on vectors). But if you have to talk in terms of angular kinetics rather than linear motion, what happens? Can you get Newton spinning?

Chapter 10 explains that there are equivalents (or *analogs*) for linear equations in angular kinetics. So, what's the angular analog for F = ma? You may guess that F, the linear force, becomes τ, or torque, after reading about torque in Chapter 10, and you'd be on the right track. And you may also guess that *a,* linear acceleration, becomes α, angular acceleration, and you'd be right. But what about *m?* What the heck is the angular analog of mass? The answer is inertia, and you come to this answer by converting tangential acceleration to angular acceleration. Your final conclusion is Στ = Iα, the angular form of Newton's second law. But I'm getting ahead of myself.

You can start the linear-to-angular conversion process with a simple example. Say that you're whirling a ball in a circle on the end of a string, as shown in Figure 11-1. And say that you apply a tangential force (along the circle) to the ball, making it speed up (keep in mind that this is a tangential force, not one directed toward the center of the circle, as when you have a centripetal force; see Chapter 10).

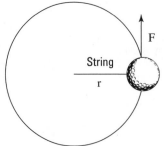

Figure 11-1: Tangential force applied to a ball on a string moving in a circle.

Start by saying that

$$F = ma$$

To convert this into terms of angular quantities like torque, multiply by the radius of the circle, *r:*

$$Fr = mra$$

Because you're applying tangential force to the ball in this case, the force and the circle's radius are at right angles (see Figure 11-1), so you can say that Fr equals torque:

$$Fr = \tau = mra$$

You're now partly done making the transition to rotational motion. Instead of working with linear force, you're working with torque, which is linear force's rotational analog.

Converting tangential acceleration to angular acceleration

To move from linear motion to angular motion, you have to convert *a*, tangential acceleration, to α, angular acceleration. Great, but how do you make the conversion? If you've pored over Chapter 10, you know that angular acceleration is a force you can multiply by the radius to get the linear equivalent, which in this case is tangential acceleration:

$$a = r\alpha$$

Substituting $a = r\alpha$ in the equation for the angular equivalent of Newton's second law (see the previous section), Fr = τ = mra, gives

$$\tau = mr(r\alpha) = mr^2\alpha$$

Now you've related torque to angular acceleration, which is what you need to go from linear motion to angular motion. But what's that mr^2 in the equation? It's the rotational analog of mass, officially called the *moment of inertia.*

Factoring in the moment of inertia

To go from linear force, F = ma, to torque (linear force's angular equivalent), you have to find the angular equivalent of acceleration and mass. In the previous section, you find angular acceleration. In this section, you find the rotational analog for mass, known as the moment of inertia: mr^2. In physics, the symbol for inertia is *I*, so you can write the equation for angular acceleration as follows:

$$\Sigma\tau = I\alpha$$

Σ means "sum of," so $\Sigma\tau$, therefore, means net torque. The units of moment of inertia are kg-m^2. Note how close this equation is to the equation for net force:

$$\Sigma F = ma$$

$\Sigma\tau = I\alpha$ is the angular form of Newton's second law for rotating bodies: net torque equals moment of inertia multiplied by angular acceleration.

Now you can put the equation to work. Say, for example, that you're whirling the 45-g ball from Figure 11-1 in a 1.0-meter circle, and you want to speed it up by 2π radians per second^2, the official units of angular acceleration. What kind of torque do you need? You know that

$$\tau = I\alpha$$

You can drop the symbol Σ from the equation when you're dealing with only one torque, meaning the "sum of" the torques is the only torque you're dealing with.

The moment of inertia equals mr^2, so

$$\tau = I\alpha = mr^2\alpha$$

Plugging in the numbers and using the meters-kilograms-seconds (MKS) system gives you

$$\tau = I\alpha = mr^2\alpha = (0.045)(1.0)^2(2\pi) = 9\pi \times 10^{-2}\ \text{N-m}$$

Your answer, $9\pi \times 10^{-2}$ N-m, is about 0.28 Newton-meters of torque. Solving for the torque required in angular motion is much like being given a mass and a required acceleration and solving for the needed force in linear motion.

Examining Moments of Inertia

Calculating moments of inertia is fairly simple if you only have to examine the motion of spherical objects, like golf balls, that have a consistent radius. For the golf ball, the moment of inertia depends on the radius of the circle it's spinning in:

$$I = mr^2$$

Here, r is the radius at which all the mass of the golf ball is concentrated. Crunching the numbers can get a little sticky when you enter the non-golf ball world, however, because you may not be sure of what radius you should use. For example, what if you're spinning a rod around? All the mass of the rod isn't concentrated at a single radius. The problem you encounter is that when you have an extended object, like a rod, each bit of mass is at a different radius. You don't have an easy way to deal with this, so you have to sum up the contribution of each particle of mass at each different radius like this:

$$I = \Sigma mr^2$$

Therefore, if you have a golf ball at radius r_1 and another at r_2, the total moment of inertia is

$$I = \Sigma mr^2 = m(r_1^2 + r_2^2)$$

So, how do you find the moment of inertia of, say, a disk rotating around an axis stuck through its center? You have to break the disk up into tiny balls and add them all up. Trusty physicists have already completed this task for many standard shapes; I provide a list of objects you're likely to encounter, and their moments of inertia, in Table 11-1.

Table 11-1	Advanced Moments of Inertia
Object	*Moment of Inertia*
Disk rotating around its center (like a tire)	$I = (\frac{1}{2})\, mr^2$
Hollow cylinder rotating around its center (like a tire)	$I = mr^2$
Hollow sphere	$I = (\frac{2}{3})\, mr^2$
Hoop rotating around its center (like a Ferris wheel)	$I = mr^2$
Point mass rotating at radius r	$I = mr^2$
Rectangle rotating around an axis along one edge	$I = (\frac{1}{3})\, mr^2$
Rectangle rotating around an axis parallel to one edge and passing through the center	$I = (\frac{1}{12})\, mr^2$
Rod rotating around an axis perpendicular to it and through its center	$I = (\frac{1}{12})\, mr^2$
Rod rotating around an axis perpendicular to it and through one end	$I = (\frac{1}{3})\, mr^2$
Solid cylinder	$I = (\frac{1}{2})\, mr^2$
Solid sphere	$I = (\frac{2}{5})\, mr^2$

Check out the following examples to see advanced moments of inertia in action.

CD players and torque: An inertia example

Here's an interesting fact about compact disc players: They actually make the CD rotate at different angular speeds to keep the section of the CD under the laser head moving at constant linear speed. For example, say that a CD has a mass of 30 grams and a diameter of 12 centimeters, and that it starts at 700 revolutions per second when you first hit play and winds down to about 200 revolutions per second at the end of the CD 50 minutes later. What's the average torque needed to create this deceleration? You start with the torque equation:

$$\tau = I\alpha$$

From Table 11-1, you know that

$$I = (\frac{1}{2})mr^2$$

because the CD is a disk rotating around its center. The diameter of the CD is 12 cm, so the radius is 6 cm. Putting in the numbers gives you

$$I = (\tfrac{1}{2})mr^2 = (\tfrac{1}{2})(0.030)(0.06)^2 = 5.4 \times 10^{-5}\ \text{kg-m}^2$$

How about the angular acceleration, α? In Chapter 10, I give you the angular equivalent of the equation for linear acceleration:

$$\alpha = \Delta\omega\ /\ \Delta t$$

For this problem, you need to find the final angular speed minus the initial angular speed to get the angular acceleration:

$$\alpha = (\omega_f - \omega_o)\ /\ \Delta t$$

You can plug in the number for time:

$$\Delta t = 50\ \text{minutes} = 1{,}500\ \text{seconds}$$

So, what about ω_f and ω_o? You know that the initial angular velocity is 700 revolutions per second, so in terms of radians per second, you get

$$\omega_o = 700\ \text{revolutions per second} \times 2\pi\ \text{radians per revolution} = 1{,}400\pi\ \text{radians per second}$$

Similarly, you can get the final angular velocity this way:

$$\omega_f = 200\ \text{revolutions per second} \times 2\pi\ \text{radians per revolution} = 400\pi\ \text{radians per second}$$

Now you can plug in the numbers:

$$\alpha = \Delta\omega\ /\ \Delta t = (\omega_f - \omega_o)\ /\ \Delta t = (400\pi - 1{,}400\pi)\ /\ 1{,}500 = -1{,}000\pi\ /\ 1{,}500 = -2.09\ \text{rad per second}^2$$

You've found the moment of inertia and the angular acceleration, so now you can find the torque:

$$\tau = I\alpha = (5.4 \times 10^{-5})(-2.09) = -1.13 \times 10^{-4}\ \text{N-m}$$

The average torque is -1.13×10^{-4} N-m. How much force is this when applied to the outer edge — that is, at a 6-cm radius? Torque = force times the radius, so

$$F = \tau\ /\ r = -1.13 \times 10^{-4}\ /\ 0.06 = 1.88 \times 10^{-3}\ \text{N}$$

This converts to about 4.23×10^{-4} pounds, or about 6.77×10^{-4} ounces of force. It doesn't take much to slow down the CD.

Angular acceleration and torque: Another inertia example

You may not always look at an object in motion and think "angular motion," like you do when looking at a spinning CD. Take someone lifting up an object with a rope on a pulley system, for example. The rope and the object are moving in a linear fashion, but the pulley has angular motion. Here's an example to help you get a handle on how fast objects will spin up (or down) when you apply a torque to them. Say that you're pulling a 16-kg mass vertically, using a pulley of mass 1 kg and radius 10 cm, as shown in Figure 11-2. You're pulling with a force of 200N; what's the angular acceleration of the pulley?

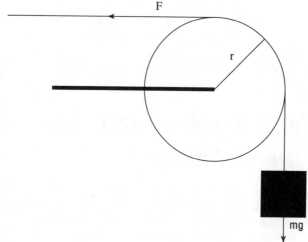

Figure 11-2: You use the torque you apply and the angular motion of the pulley to lift objects in a pulley system.

You use the equation for torque, including the sum symbol, Σ, because you're dealing with more than one torque in this problem (you always use the net torque in a problem, but many problems only have one torque, so the symbol drops off):

$$\Sigma\tau = I\alpha$$

where $\Sigma\tau$ means net torque. In this case, there are two torques, τ_1 and τ_2. So,

$$\alpha = \Sigma\tau \,/\, I = (\tau_1 + \tau_2) \,/\, I$$

The two forces act at radius 10.0 cm, so the two torques are

$\tau_1 = Fr$, with F as the force and r as the radius

$\tau_2 = -mgr$, with m as the mass and g as the acceleration due to gravity

The pulley's support goes through the axis of rotation, so no torque comes from it. Plugging in the numbers gives you

$$\tau_1 = Fr = (200)(0.10) = 20 \text{ N-m}$$

$$\tau_2 = -mgr = -(16)(9.8)(0.10) = -15.68 \text{ N-m}$$

The net torque is $20 + -15.68 = 4.32$ N-m. The result is positive, so the pulley will rotate in a counterclockwise direction, toward larger angles, which means the mass will rise.

To find the angular acceleration of the pulley, you need to know its moment of inertia. From Table 11-1, you know that for a disk like a pulley, the moment is $I = (\frac{1}{2})mr^2$, so

$$I = (\tfrac{1}{2})mr^2 = (\tfrac{1}{2})[(1.0)(0.10)^2] = 5.0 \times 10^{-3} \text{ kg-m}^2$$

You have all you need to find the angular acceleration:

$$\alpha = \Sigma\tau / I = 4.32 / 5.0 \times 10^{-3} = 864 \text{ radians per second}^2$$

Wrapping Your Head around Rotational Work and Kinetic Energy

One major player in the linear-force game is *work* (see Chapter 8); the equation for work is work = force times distance. Work has a rotational analog — but how the heck can you relate a linear force acting for a certain distance with the idea of rotational work? You convert force to torque, its angular equivalent, and distance to angle. I show you the way in the following sections, and I show you what happens when you do work by turning an object, creating rotational motion — the same thing that happens when you do work in linear motion: You produce energy.

Doing some rotational work

Imagine that an automobile engineer is sitting around contemplating a new, fresh idea in car design. What she wants to create is something environmentally responsible *and* daring — something never before seen in the industry.

In a burst of inspiration, the answer comes to her — not only the answer to her automobile dreams, but also your answer on how to tie linear work to rotational work. Take a look at Figure 11-3, where I've sketched out the whole solution. What she should do is tie a string around the car tire so the driver can simply pull the string to accelerate the car! This allows her to give feedback to the drivers on how much work they're doing.

String tied around tire

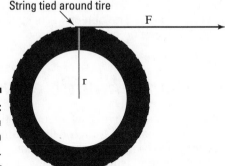

F

r

Figure 11-3: Exerting a force to turn a tire.

Work is the amount of force applied to an object multiplied by the distance it's applied. In this case, the drivers apply a force F, and they apply that force with the string. Bingo! The string is what lets you make the handy transition between linear and rotational work. So, how much work is done? Use the equation

$$W = Fs$$

where s is the distance the driver applies the force over. In this case, the distance s equals the radius multiplied by the angle through which the wheel turns, $s = r\theta$, so you get

$$W = Fs = Fr\theta$$

However, the torque, τ, equals Fr in this case, because the string is acting at right angles to the radius (see Chapter 10). So you're left with

$$W = Fs = Fr\theta = \tau\theta$$

The work done, W, by turning the wheel with a string and a constant torque is $\tau\theta$. This makes sense, because linear work is Fs, and to convert to rotational work, you convert from force to torque and from distance to angle. The units here are the standard units for work — Joules in the MKS system, for instance.

Note that you have to give the angle in radians for the conversion between linear work and rotational work to come out right.

For example, say that you have a plane that uses propellers to fly, and you want to determine how much work the plane's engine does on a propeller when applying a constant torque of 600 N-m over 100 revolutions. You start with

$$W = \tau\theta$$

A full revolution is 2π radians, so plugging the numbers into the equation gives you

$$W = \tau\theta = (600)(100 \times 2\pi) = 3.77 \times 10^5 \text{ J}$$

The plane's engine does 3.77×10^5 Joules of work. But what happens when you put a lot of work into turning an object? The object starts spinning. And, just as with linear work, the work you do becomes *energy*.

Tracking down rotational kinetic energy

When an object is spinning, all its pieces are moving, which means that kinetic energy is at work. However, you have to convert from the linear concept of kinetic energy to the rotational concept of kinetic energy. You can calculate the kinetic energy of a body in linear motion with the following equation (see Chapter 8):

$$KE = \tfrac{1}{2} mv^2$$

where m is the mass of the object and v^2 is the square of its speed. This applies to every bit of the object that's rotating — each bit of mass has this kinetic energy.

To go from the linear version to the rotational version, you have to go from mass to moment of inertia, I, and from velocity to angular velocity, ω.

You can tie an object's tangential speed to its angular speed like this (see Chapter 10):

$$v = r\omega$$

where r is the radius and ω is its angular speed. Plugging this into the previous equation gives you

$$KE = \tfrac{1}{2} mv^2 = \tfrac{1}{2} m(r^2\omega^2)$$

The equation looks okay so far, but remember that it holds true only for the one single bit of mass under discussion — each other bit of mass may have a different radius, for example, so you're not finished. You have to sum up the kinetic energy of every bit of mass like this:

$$KE = \tfrac{1}{2} \Sigma(mr^2\omega^2)$$

You may be wondering if you can simplify this equation. Well, you can start by noticing that even though each bit of mass may be different and be at a different radius, each bit has the same angular speed (they all turn through the same angle in the same time). Therefore, you can take the ω out of the summation:

$$KE = \tfrac{1}{2} \, \Sigma(mr^2) \, \omega^2$$

This makes the equation much simpler, because the moment of inertia, I (see the section "Rolling Up Newton's Second Law into Angular Motion" earlier in this chapter), equals $\Sigma(mr^2)$. Making this substitution takes all the dependencies on the individual radius of each bit of mass out of the equation, giving you

$$KE = \tfrac{1}{2} \, \Sigma(mr^2) \, \omega^2 = \tfrac{1}{2} \, I\omega^2$$

Now you have a simplified equation for rotational kinetic energy. The equation proves useful, because rotational kinetic energy is everywhere. A satellite spinning around in space has rotational kinetic energy. A barrel of beer rolling down a ramp from a truck has rotational kinetic energy. The latter example (not always with beer trucks, of course) is a common thread in physics problems.

Measuring rotational kinetic energy on a ramp

Objects can have both linear and rotational kinetic energy. This fact is an important one, if you think about it, because when objects start rolling down ramps, any previous ramp expertise you have goes out the window. Why? Because when an object rolls down a ramp instead of sliding, some of its gravitational potential energy (see Chapter 8) goes into its linear kinetic energy, and some of it goes into its rotational kinetic energy.

Take a look at Figure 11-4, where you're pitting a solid cylinder against a hollow cylinder in a race down the ramp. Each object has the same mass. Which cylinder is going to win?

You can rephrase this question as: Which cylinder will have the higher speed at the bottom of the ramp? The acceleration of each cylinder will be constant, so the one with the higher final speed will be the one that traveled the quickest. When looking only at linear motion, you can handle a problem like this by setting the final kinetic energy equal to the potential energy (assuming no friction!) like this:

$$KE = PE \text{ (potential energy)} = \tfrac{1}{2} \, mv^2 = mgh$$

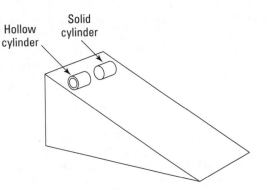

Figure 11-4:
A solid
cylinder and
a hollow
cylinder
preparing
for a race
down a
ramp.

where m is the mass of the object, g is the acceleration due to gravity, and h is its height. This would let you solve for the final speed. But the situation is different now because the cylinders are rolling, which means that the initial gravitational potential energy becomes both linear kinetic energy *and* rotational kinetic energy. You can now write the equation as

$$mgh = \tfrac{1}{2}\,mv^2 + \tfrac{1}{2}\,I\omega^2$$

You can relate v and ω together with the equation $v = r\omega$, which means that $\omega = v\,/\,r$, so

$$mgh = \tfrac{1}{2}\,mv^2 + \tfrac{1}{2}\,I\omega^2 = \tfrac{1}{2}\,mv^2 + \tfrac{1}{2}\,I(v^2\,/\,r^2)$$

You want to get v out of the equation, so try grouping things together:

$$mgh = \tfrac{1}{2}\,mv^2 + \tfrac{1}{2}\,I\omega^2 = \tfrac{1}{2}\,mv^2 + \tfrac{1}{2}\,I(v^2\,/\,r^2) = \tfrac{1}{2}(m + I\,/\,r^2)v^2$$

Isolating v, you get

$$v = \sqrt{\frac{2mgh}{m + I/r^2}}$$

What's this break down to when you substitute for the two different moments of inertia? For a hollow cylinder, the moment of inertia equals mr^2, as you can see in Table 11-1. For a solid cylinder, on the other hand, the moment of inertia equals $(\tfrac{1}{2})\,mr^2$. Substituting for I for the hollow cylinder gives you

$$v = \sqrt{gh}$$

Substituting for I for the solid cylinder gives you

$$v = \sqrt{\frac{4gh}{3}}$$

Now the answer becomes clear. The solid cylinder will be going

$$\sqrt{\frac{4}{3}}$$

times as fast as the hollow cylinder, or about 1.15 times as fast, so the solid cylinder will win.

Does it make sense intuitively that the solid cylinder in this section wins the race? It does, because the hollow cylinder has as much mass concentrated at a large radius as the solid cylinder has distributed from the center all the way out to that radius. With that large mass way out at the edge, the hollow cylinder won't need to go as fast to have as much rotational kinetic energy as the solid cylinder.

Can't Stop This: Angular Momentum

Picture a 40-ton satellite rotating in orbit around the earth. You may want to stop it to perform some maintenance, but when it comes time to grab it, you stop and consider the situation. It takes a lot of effort to stop that spinning satellite. Why? Because it has *angular momentum.*

In Chapter 9, I cover linear momentum, *p,* which equals the product of mass and velocity:

p = mv

Physics also features angular momentum. Its letter, L, has as little to do with the word "momentum" as the letter *p* does. The equation for angular momentum looks like this:

L = Iω

where *I* is the moment of inertia and ω is the angular velocity.

Note that angular momentum is a vector quantity, meaning it has a magnitude and a direction, that points in the same direction as the ω vector (that is, in the direction the thumb of your right hand points when you wrap your fingers around in the direction the object is turning).

The units of angular momentum are *I* multiplied by ω, or kg-m^2/s in the MKS system (see Chapter 2 for more on measurement systems).

The important fact about angular momentum, much as with linear momentum, is that it's conserved.

Reviewing the conservation of angular momentum

The principle of conservation of angular momentum states that angular momentum is conserved if there are no net torques involved. This principle comes in handy in all sorts of problems, often where you least expect it. You may come across more obvious cases, like when two ice skaters start off holding each other close while spinning but then end up at arm's length. Given their initial angular speed, you can find their final angular speed, because angular momentum is conserved, which tells you the following is true:

$$I_1\omega_1 = I_2\omega_2$$

If you can find the initial moment of inertia and the final moment of inertia, you're set. But you also come across less obvious cases where the principle of conservation of angular momentum helps out. For example, satellites don't have to travel in circular orbits; they can travel in ellipses. And when they do, the math can get a lot more complicated. Lucky for you, the principle of conservation of angular momentum can make the problems childishly simple.

Satellite orbits: A conservation of angular momentum example

Say, for example, that NASA planned to put a satellite into a circular orbit around Pluto for studies, but the situation got a little out of hand, and the satellite ended up with an elliptical orbit. At its nearest point to Pluto, 6.0×10^6 meters, the satellite zips along at 9,000 meters per second.

The satellite's farthest point from Pluto is 2.0×10^7 meters. What's its speed at that point? The answer is tough to figure out unless you can come up with an angle here, and that angle is angular momentum. Angular momentum is conserved here, because there are no external torques the satellite must deal with (gravity always acts perpendicular to the orbital radius). Because angular momentum is conserved, you can say that

$$I_1\omega_1 = I_2\omega_2$$

For a point mass like a satellite (it's a point mass compared to the radius of its orbit at any location), the moment of inertia, I, equals mr^2 (see Table 11-1). The angular velocity equals v / r, so the equation works out to be

$$I_1\omega_1 = I_2\omega_2 = mr_1v_1 = mr_2v_2$$

You can put v_2 on one side of the equation by dividing by mr_2:

$$v_2 = (r_1v_1) / r_2$$

You have your solution; no fancy math involved at all, because you can rely on the principle of conservation of angular momentum to do the work for you. All you need to do is plug in the numbers:

$$v_2 = (r_1v_1) / r_2 = [(6.0 \times 10^6)(9,000)] / 2.0 \times 10^7 = 2,700 \text{ meters per second}$$

At its closest point to Pluto, the satellite will be screaming around at 9,000 meters per second, and at its farthest point, it will be moving at 2,700 meters per second. Easy enough to figure out, as long as you have the principle of conservation of angular momentum under your belt.

Chapter 12

Springs-n-Things: Simple Harmonic Motion

*I*n this chapter, I shake things up with a new kind of motion: periodic motion, which occurs when objects are bouncing around on springs or bungee cords or are even swooping around on the end of a pendulum. This chapter is all about describing their periodic motion. Not only can you describe their motions in detail, but you can also predict how much energy bunched-up springs have, how long it will take a pendulum to swing back and forth, and more.

Hooking Up with Hooke's Law

Objects, like springs, that you can stretch but that return to their original shapes are called *elastic*. Elasticity is a valuable property — it means you can use springs for all kinds of applications: as shock absorbers in lunar landing modules, as timekeepers in clocks and watches, and even as hammers of justice in mousetraps.

As long ago as the 1600s, Robert Hooke, a physicist from England, undertook the study of elastic materials. He created a new law, not surprisingly called *Hooke's law,* which states that stretching an elastic material gives you a force that's directly proportional to the amount of stretching you do. For example, if you stretch a spring a distance x, you'll get a force back that's directly proportional to x:

$$F = kx$$

where k is the spring constant. In fact, the force F resists your pull, so it pulls in the opposite direction, which means you should have a negative sign here:

$$F = -kx$$

Keeping springs stretchy

Hooke's law is valid as long as the elastic material you're dealing with stays elastic — that is, it stays within its *elastic limit.* If you pull a spring too far, it loses its stretchy ability, for example. In other words, as long as a spring stays within its elastic limit, you can say that $F = -kx$, where the constant k is called the *spring constant.* The constant's units are Newtons per meter. When a spring stays within its elastic limit, it's called an *ideal spring.*

Say, for example, that a group of car designers knocks on your door and asks if you can help design a suspension system. "Sure," you say. They inform you that the car will have a mass of 1,000 kg, and you have four shock absorbers, each 0.5 meters long, to work with. How strong do the springs have to be? Assuming these shock absorbers use springs, each one has to support, at a very minimum, a weight of 250 kg, which is

$$F = mg = (250)(9.8) = 2{,}450 \text{ Newtons}$$

where F equals force, m equals the mass of the object, and g equals the acceleration due to gravity, 9.8 meters per second^2. The spring in the shock absorber will, at a minimum, have to give you 2,450N of force at the maximum compression of 0.5 meters. What does this mean the spring constant should be? Hooke's law says

$$F = -kx$$

Omitting the negative sign (look for its return in the following section), you get

$$k = F / x$$

Time to plug in the numbers:

$$k = F / x = 2{,}450N / 0.5 \text{ m} = 4{,}900 \text{ Newtons/meter}$$

The springs used in the shock absorbers must have spring constants of at least 4,900 Newtons per meter. The car designers rush out, ecstatic, but you call after them: "Don't forget, you need to at least double that in case you actually want your car to be able to handle pot holes . . ."

Deducing that Hooke's law is a restoring force

The negative sign in Hooke's law for an elastic spring is important:

$$F = -kx$$

The negative sign means that the force will oppose your displacement, as you see in Figure 12-1, which shows a ball attached to a spring.

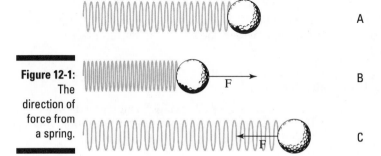

Figure 12-1: The direction of force from a spring.

A

B

C

As you see in Figure 12-1, if the spring isn't stretched or compressed, it exerts no force on the ball. If you push the spring, however, it pushes back, and if you pull the spring, it pulls back.

The force exerted by a spring is called a *restoring force.* It always acts to restore itself toward equilibrium.

Moving with Simple Harmonic Motion

An object undergoes *simple harmonic motion* when the force that tries to restore the object to its rest position is proportional to the displacement of the object. Simple harmonic motion is simple because the forces involved are elastic, which means you can assume that no friction is involved. Elastic forces insinuate that the motion will just keep repeating. That isn't really true, however; even objects on springs quiet down after a while as friction and heat loss in the spring take their toll. The harmonic part of simple harmonic motion means that the motion repeats — just like harmony in music, where vibrations create the sound you hear.

Examining basic horizontal and vertical simple harmonic motion

Take a look at the ball in Figure 12-1. Say, for example that you push the ball, compressing the spring, and then you let go; the ball will shoot out, stretching the spring. After the stretch, the spring will pull back, and it will once again pass the equilibrium point (where no force acts on the ball), shooting backward past it. This event happens because the ball has inertia (see Chapter 11), and when it's moving, it takes some force to bring it to a stop. Here are the various stages the ball goes through, matching the letters in Figure 12-1 (and assuming no friction):

- ✔ **Point A:** The ball is at equilibrium, and no force is acting on it. This is called the *equilibrium point,* where the spring isn't stretched or compressed.

- ✔ **Point B:** The ball pushes against the spring, and the spring retaliates with force F opposing that pushing.

- ✔ **Point C:** The spring releases, and the ball springs to an equal distance on the other side of the equilibrium point. At this point, it isn't moving, but a force acts on it, F, so it starts going back the other direction.

The ball passes through the equilibrium point on its way back to point B. At this equilibrium point, the spring doesn't exert any force on the ball, but the ball is traveling at its maximum speed. This is what happens when the golf ball bounces back and forth; you push the ball to point B, it goes through point A, moves to point C, shoots back to A, moves to B, and so on: B-A-C-A-B-A-C-A, and so on. Point A is the equilibrium point, and both points B and C are equidistant from point A.

What if the ball in Figure 12-1 wasn't on a frictionless horizontal surface? What if it were to hang on the end of a spring in the air, as shown in Figure 12-2?

In this case, the ball oscillates up and down. Like when the ball was on a surface in Figure 12-1, it will oscillate around the equilibrium position; this time, however, the equilibrium position isn't the point where the spring isn't stretched.

The equilibrium position is defined as the position at which no net force acts on the ball. In other words, the equilibrium position is the point where the ball is at rest. When the spring is held vertically, the weight of the ball downward is matched by the pull of the spring upward. If the y position of the ball corresponds to the equilibrium point, y_o, because the weight of the ball, *mg*, must match the force exerted by the spring, $F = ky_o$, you have

$$mg = ky_o$$

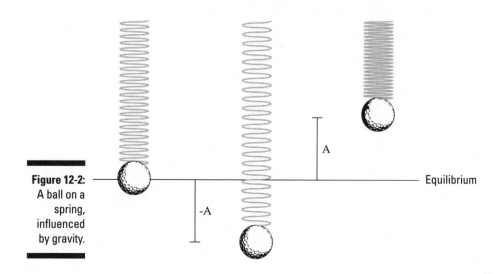

Figure 12-2:
A ball on a
spring,
influenced
by gravity.

Equilibrium

A

-A

And you can solve for y_o:

$$y_o = mg / k$$

This equation represents the distance the spring will stretch because the ball is attached to it. When you pull the ball down or lift it up and then let go, it oscillates around the equilibrium position, as shown in Figure 12-2. If the spring is completely elastic, the ball will undergo simple harmonic motion vertically around the equilibrium position; the ball goes up a distance A and down a distance A around that position (in real life, the ball would eventually come to rest at the equilibrium position).

The distance A, or how high the object springs up, is an important one when describing simple harmonic motion; it's called the *amplitude.* You can describe simple harmonic motion pretty easily by using some math, and the amplitude is an important part of that.

Diving deeper into simple harmonic motion

Calculating simple harmonic motion can require time and patience when you have to figure out how the motion of an object changes over time. Imagine that one day you come up with a brilliant idea for an experimental apparatus. You decide that a spotlight would cast a shadow of a ball on a moving piece of photographic film, as you see in Figure 12-3. Because the film is moving, you get a record of the ball's motion as time goes on.

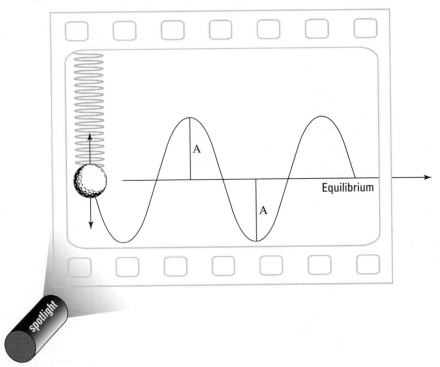

You turn the apparatus on and let it do its thing. The results are shown in Figure 12-3 — the ball oscillates around the equilibrium position, up and down, reaching amplitude A at its lowest and highest points. But take a look at the ball's track. Near the equilibrium point, it goes its fastest, because a lot of force accelerates it. At the top and bottom, it's subject to plenty of force, so it slows down and reverses its motion.

You deduce that the track of the ball is a *sine wave,* which means that its track is a sine wave of amplitude A. You can also use a cosine wave, because the shape is the same. The only difference is that when a sine wave is at its peak, the cosine wave is at zero, and vice versa.

Breaking down the sine wave

You can get a clear picture of the sine wave if you plot the sine function on an X-Y graph like this:

$$y = \sin(x)$$

You see the kind of shape you see in Figure 12-3 on the plot — a sine wave — which may already be familiar to you from your explorations of math or from other places, such as the screens of heart monitors in movies or on television. Take a look at the sine wave in a circular way. If the ball from Figure 12-3 were attached to a rotating disk, as you see in Figure 12-4, and you shine a spotlight on it, you'd get the same result as when you have the ball hanging from the spring: a sine wave.

The rotating disk is often called a *reference circle,* which you can see in Figure 12-5. In the figure, the view is from above, and the ball is glued to the turning disk. Reference circles can tell you a lot about simple harmonic motion.

As the disk turns, the angle, θ, increases in time. What will the track of the ball on the film look like as the film moves up out of the page? You can resolve the motion of the ball yourself along the X-axis; all you need is the X component of the ball's motion. At any one time, the ball's X position will be

x = A cos θ

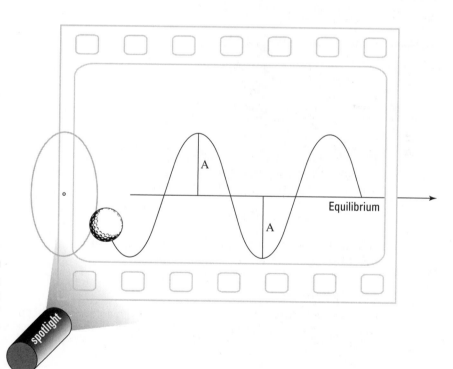

Figure 12-4:
An object
with circular
motion
becomes a
sine wave.

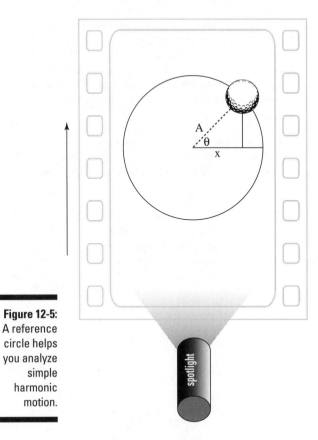

Figure 12-5:
A reference
circle helps
you analyze
simple
harmonic
motion.

This varies from positive A to negative A in amplitude. In fact, you can say that you already know how θ is going to change in time, because $\theta = \omega \times t$, where ω is the angular speed of the disk and t is time:

$$x = A \cos \theta = A \cos(\omega t)$$

You can now explain what the track of the ball will be as time goes on, given that the disk is rotating with angular speed ω.

Getting periodic

Each time an object moves around a full circle, it completes a *cycle*. The time it takes to complete the cycle is called the *period*. The letter used for period is T, and it's measured in seconds.

Looking at Figure 12-5, in terms of the x motion on the film, during one cycle, the ball moves from x = A to –A and then back to A. In fact, when it goes from any point on the sine wave (see the previous section) back to the same point on the sine wave later in time, it completes a cycle. And the time the ball takes to move from a certain position back to that same position while moving in the same direction is its period.

How can you relate the period to something more familiar? When an object moves around in a circle, it goes 2π radians. It travels that many radians in T seconds, so its angular speed, ω (see Chapter 10), is

$$\omega = 2\pi / T$$

Multiplying both sides by T and dividing by ω allows you to relate the period and the angular speed:

$$T = 2\pi / \omega$$

Sometimes, however, you speak in terms of the frequency of periodic motion, not the period. The *frequency* is the number of cycles per second. For instance, if the disk from Figure 12-4 rotates at 1,000 full turns per second, the frequency, *f*, would be 1,000 cycles per second. Cycles per second are also called *Hertz*, abbreviated Hz, so this would be 1,000 Hz.

So, how do you connect frequency, *f*, to period, T? T is the amount of time one cycle takes, so you get

$$f = 1 / T$$

You've found that $\omega = 2\pi / T$, so you can modify this equation to get the following:

$$\omega = (2\pi / T) \times Tf = 2\pi f$$

So far, you've known ω only as the angular speed. But when you're dealing with springs, you don't have a lot of angles involved, so you call ω the *angular frequency* instead.

Remembering not to speed away without the velocity

Take a look at Figure 12-5, where a ball is rotating on a disk. In the section "Breaking down the sine wave," earlier in this chapter, you figure out that

$$x = A \cos (\omega t)$$

where *x* stands for the X coordinate and A stands for the amplitude of the motion. But in that section, you don't realize that other forces are at work. At any point X, the ball also has a certain velocity, which varies in time also. So, how can you describe the velocity mathematically? Well, you can relate tangential velocity to angular velocity like this (see Chapter 10):

$$v = r\omega$$

where r represents the radius. Because $r = A$, you get

$$v = r\omega = A\omega$$

Does this equation get you anywhere? Sure, because the shadow of the ball on the film gives you simple harmonic motion. The velocity vector (the direction of the magnitude of the velocity; see Chapter 4) always points tangentially here, perpendicular to the radius, so you get the following for the X component of the velocity at any one time:

$$v_x = -A\omega \sin \theta$$

The negative sign here is important, because the X component of the velocity of the ball in Figure 12-5 points to the left, toward negative X. And because the ball is on a rotating disc, you know that $\theta = \omega t$, so

$$v_x = -A\omega \sin \theta = -A\omega \sin (\omega t)$$

This equation describes the velocity of a ball in simple harmonic motion. Note that the velocity changes in time — from $-A\omega$ to 0 and then to $A\omega$ and back again. So, the maximum velocity, which happens at the equilibrium point, has a magnitude of $A\omega$. This says, among other things, that the velocity is directly proportional to the amplitude of the motion; as amplitude increases, so does velocity, and vice versa.

For example, say that you're on a physics expedition watching a daredevil team do some bungee jumping. You notice that the team members are starting by finding the equilibrium point of their new bungee cords, so you measure that point.

The team decides to let their leader go a few meters above the equilibrium point, and you watch as he flashes past the point and then bounces back at a speed of 4.0 meters per second as of the equilibrium point. Ignoring all caution, the team lifts its leader to a distance 10 times greater away from the equilibrium point and lets go again. This time you hear a distant scream as the costumed figure hurtles up and down. What's his maximum speed?

You know that he was going 4.0 meters per second at the equilibrium point before, the point where he achieves maximum speed; you know that he started with an amplitude 10 times greater on the second try; and you know that the maximum velocity is proportional to the amplitude. Therefore, assuming that the frequency of his bounce is the same, he'll be going 40.0 meters per second at the equilibrium point — pretty speedy.

Including the acceleration

You can find the displacement of an object undergoing simple harmonic motion with the following equation:

$$x = A \cos (\omega t)$$

And you can find the object's velocity with the equation

$$v = -A\omega \sin(\omega t)$$

But you have another factor to account for when describing an object in simple harmonic motion: its acceleration at any particular point. How do you figure it out? No sweat. When an object is going around in a circle, the acceleration is the centripetal acceleration (see Chapter 10), which is

$$a = r\omega^2$$

where r is the radius and ω is the angular speed. And because $r = A$ — the amplitude — you get

$$a = r\omega^2 = A\omega^2$$

This equation represents the magnitude of the centripetal acceleration. To go from a reference circle (see the section "Breaking down the sine wave" earlier in this chapter) to simple harmonic motion, you take the component of the acceleration in one dimension — the X direction here — which looks like

$$a = -A\omega^2 \cos\theta$$

The negative sign indicates that the X component of the acceleration is toward the left. And because $\theta = \omega t$, where t represents time, you get

$$a = -A\omega^2 \cos\theta = -A\omega^2 \cos(\omega t)$$

Now you have the equation to find the acceleration of an object at any point while it's moving in simple harmonic motion. For example, say that your phone rings, and you pick it up. You hear "Hello?" from the earpiece.

"Hmm," you think. "I wonder what g forces (forces exerted on an object due to gravity) the diaphragm in the phone is undergoing."

The diaphragm (a metal disk that acts like an eardrum) in your phone undergoes a motion very similar to simple harmonic motion, so calculating its acceleration isn't any problem. You pull out your calculator; measuring carefully, you note that the amplitude of the diaphragm's motion is about 1.0×10^{-4}m. So far, so good. Human speech is in the 1.0-kHz (kilohertz, or 1,000 Hz) frequency range, so you have the frequency, ω. And you know that

$$a_{max} = A\omega^2$$

Also, $\omega = 2\pi f$, where f represents frequency, so plugging in the numbers gives you

$$a_{max} = A\omega^2 = (1.0 \times 10^{-4})[(2\pi)(1,000)]^2 = 3,940 \text{ m/s}^2$$

You get a value of 3,940 meters per second^2, which is about 402 g.

"Wow," you say. "That's an incredible acceleration to pack into such a small piece of hardware."

"What?" says the impatient person on the phone. "Are you doing physics again?"

Finding the angular frequency of a mass on a spring

If you take the information you know about Hooke's law for springs (see the section "Hooking Up with Hooke's Law" earlier in this chapter) and apply it to what you know about finding simple harmonic motion (see the section "Moving with Simple Harmonic Motion"), you can find the angular frequencies of masses on springs, along with the frequencies and periods of oscillations. And because you can relate angular frequency and the masses on springs, you can find the displacement, velocity, and acceleration of the masses.

Hooke's law says that

$$F = -kx$$

where F is the force, k is the spring constant, and x is distance. Because of Newton (see Chapter 5), you also know that Force = mass times acceleration, so you get

$$F = ma = -kx$$

This equation is in terms of displacement and acceleration, which you see in simple harmonic motion in the following forms (see the previous section in this chapter):

$$x = A \cos(\omega t)$$
$$a = -A\omega^2 \cos(\omega t)$$

Inserting these two equations into the previous equation gives you

$$F = ma = -mA\omega^2 \cos(\omega t) = -kx = -kA \cos(\omega t)$$

This equation breaks down to

$$m\omega^2 = k$$

Rearranging to put ω on one side gives you

$$\omega = \sqrt{\frac{k}{m}}$$

You can now find the angular frequency for a mass on a spring, and it's tied to the spring constant and the mass. You can also tie this to the frequency of oscillation and to the period of oscillation (see the section "Breaking down the sine wave) by using the following equation:

$$\omega = 2\pi / T = 2\pi f$$

You can convert this to

$$f = 2\pi \sqrt{\frac{k}{m}}$$

and

$$T = \frac{1}{2\pi} \sqrt{\frac{m}{k}}$$

Say, for example, that the spring in Figure 12-1 has a spring constant, k, of 1.0×10^2 Newtons per meter and that you attach a 45-g ball to the spring. What's the period of oscillation? All you have to do is plug in the numbers:

$$T = \frac{1}{2\pi} \sqrt{\frac{m}{k}} = \frac{1}{2\pi} \sqrt{\frac{0.045}{1.0 \times 10^{-2}}} = 0.338 \text{ seconds}$$

The period of the oscillation is 0.34 seconds. How many bounces will you get per second? The number of bounces represents the frequency, which you find this way:

$$f = 1/T = 2.96 \text{Hz}$$

You get about 3 oscillations per second.

Because you can tie the angular frequency, ω, to the spring constant and the mass on the end of the spring, you can predict the displacement, velocity, and acceleration of the mass, using the following equations for simple harmonic motion (see the section "Diving deeper into simple harmonic motion" earlier in this chapter):

$$x = A \cos(\omega t)$$

$$v = -A\omega \sin(\omega t)$$

$$a = -A\omega^2 \cos(\omega t)$$

Using the previous example of the spring in Figure 12-1 — having a spring constant of 1.0×10^2 Newtons per meter with a 45-g ball attached — you know that

$$\omega = \sqrt{\frac{k}{m}} = 0.471 \ \text{sec}^{-1}$$

Say, for example, that you pull the ball 10.0 cm before releasing it (making the amplitude 10.0 cm). In this case, you know that

$$x = (0.10) \cos (0.471t)$$

$$v = -(0.10)(0.471) \sin (0.471t)$$

$$a = -(0.10)(0.471)^2 \cos (0.471t)$$

Factoring Energy into Simple Harmonic Motion

Along with the actual motion that takes place (or that you cause) in simple harmonic motion, you can examine the energy involved. For example, how much energy is stored in a spring when you compress or stretch it? The work you do compressing or stretching the spring must go into the energy stored in the spring. That energy is called *elastic potential energy* and is the force, F, times the distance, *s:*

$$W = Fs$$

As you stretch or compress a spring, the force varies, but it varies in a linear way, so you can write the equation in terms of the average force, $\bar{F}$:

$$W = \bar{F}s$$

The distance (or displacement), *s*, is just the difference in position, $x_f - x_o$, and the average force is $\frac{1}{2}(F_f + F_o)$, which means

$$W = \bar{F}s = [\tfrac{1}{2}(F_f + F_o)](x_f - x_o)$$

Substituting Hooke's law (see the section "Hooking Up with Hooke's Law" earlier in this chapter), $F = -kx$, for F gives you

$$W = \bar{F}s = [\tfrac{1}{2}(F_f + F_o)](x_f - x_o) = [-\tfrac{1}{2}(kx_f + kx_o)](x_f - x_o)$$

Simplifying the equation gives you

$$W = \tfrac{1}{2}kx_o^2 - \tfrac{1}{2}kx_f^2$$

The work done on the spring changes the potential energy stored in the spring. The following is exactly how you give that potential energy, or the elastic potential energy:

$$PE = \tfrac{1}{2}kx^2$$

For example, if a spring is elastic and has a spring constant, *k,* of 1.0×10^2 Newtons per meter, and you compress it by 10.0 cm, you store the following amount of energy in it:

$$PE = \tfrac{1}{2}kx^2 = \tfrac{1}{2}(1.0 \times 10^2)(0.10)^2 = 5.0 \times 10^5 J$$

You can also note that when you let the spring go with a mass on the end of it, the mechanical energy (the sum of potential and kinetic energy) is conserved:

$$PE_1 + KE_1 = PE_2 + KE_2$$

When you compress the spring 10.0 cm, you know that you have 5.0×10^5 J of energy stored up. When the moving mass reaches the equilibrium point and no force from the spring is acting on the mass, you have maximum velocity and therefore maximum kinetic energy — at that point, the kinetic energy is 5.0×10^5 J, by the conservation of mechanical energy (see Chapter 8 for more on this topic).

Swinging with Pendulums

Other objects move in simple harmonic motion besides springs, such as the pendulum you see in Figure 12-6. Here, a ball is tied to a string and is swinging back and forth.

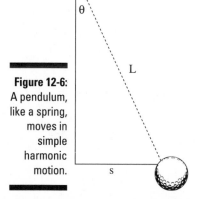

Figure 12-6:
A pendulum, like a spring, moves in simple harmonic motion.

Can you analyze this pendulum's motion as you would a spring's (see the section "Moving with Simple Harmonic Motion")? Yep, no problem. Take a look at Figure 12-6. The torque, τ (see Chapter 10), that comes from gravity is the weight of the ball, negative mg, multiplied by the lever arm, s (for more on lever arms, see Chapter 10):

$$\tau = -mgs$$

Here's where you make an approximation. For small angles θ, the distance s equals $L\theta$, where L is the length of the pendulum string:

$$\tau = -mgs = -(mgL)\theta$$

This equation resembles the form of Hooke's law, $F = -kx$ (see the section "Hooking Up with Hooke's Law"), if you treat mgL as you would a spring constant. And you can use the rotational moment of inertia, I (see Chapter 11), rather than the mass for the ball; doing so lets you solve for the angular frequency of the pendulum in much the same way you solve for the angular frequency of a spring (see the section "Finding the angular frequency of a mass on a spring" earlier in this chapter):

$$\omega = \sqrt{\frac{mgL}{I}}$$

The moment of inertia equals mr^2 for a point mass (see Chapter 11), which you can use here, assuming that the ball is small compared to the pendulum string. This gives you

$$\omega = \sqrt{\frac{g}{L}}$$

Now you can plug this into the equations of motion for simple harmonic motion. You can also find the period of a pendulum with the equation

$$\omega = 2\pi / T = 2\pi f$$

where T represents period and f represents frequency. You end up with the following for the period:

$$T = 2\pi \sqrt{\frac{L}{g}}$$

Note that this period is actually independent of the mass you're using on the pendulum!

Part IV

Laying Down the Laws of Thermodynamics

The 5th Wave   By Rich Tennant

THE 4th LAW OF THERMODYNAMICS: Don't try to explain the first 3 on a blind date.

In this part . . .

How much boiling water does it take to melt a 200-pound block of ice? Why would you freeze in space? Why does metal feel cold to the touch? It all boils (or freezes) down to thermodynamics, which is the physics of heat and heat flow. The answers to your questions are coming in this part in the form of useful equations and explanations.

Chapter 13

Turning Up the Heat with Thermodynamics

*Y*ou arrive at the scene of an emergency where an unexpected geothermal event is ruining a garden birthday party. "It's a geyser," the unhappy parents say, pointing out a foaming pit in the backyard. "It sure is," you say. "Have you got a meter stick?" The parents pass you a meter stick and say, "Could you hurry it up? The ice cream is melting."

You swiftly measure the depth of the column of boiling water at 225 meters, at an average width of 0.5 meters. "No problem," you announce. "Physics has given us the solution. You need 719 bags of ice."

"Seven hundred nineteen bags of . . . ice?" the parents echo faintly.

"The bags will cool the geyser just enough for the birthday party to finish, assuming you have exactly 120 minutes or fewer to go."

"Seven hundred nineteen bags?" the parents ask, looking at each other. "Yep," you say. "I'll send you my bill."

This chapter is an exploration of heat and temperature. Physics has plenty to say on the subjects, and it gives you plenty of power to predict what goes on. I discuss different temperature measurements, linear expansion, volume expansion, and how much of an object at one temperature will change the temperature of another object when they're put together.

Getting into Hot Water

You always start a calculation or observation in physics by making measurements; and when the physics topics you're discussing are heat and temperature, you have several different scales at your disposal — most notably, Fahrenheit, Celsius, and Kelvin.

When the thermometer says Fahrenheit

In the United States, the most common temperature scale is the *Fahrenheit scale*. Like in other temperature scales, measurements in the Fahrenheit scale are made in degrees. For example, the blood temperature of a healthy human being is 98.6°F — the F means you're using the Fahrenheit scale.

When the thermometer says Celsius

The Fahrenheit system wasn't very reproducible in its early days, so another system was developed — the *Celsius scale* (formerly called the Centigrade system). Using this system, pure water freezes at 0°C, and it boils at 100°C. These measurements are easier to reproduce than Fahrenheit's 32° and 212° because pure water is easy to come up with. Here's how you tie the two systems of temperature measurement together (these measurements are at sea level; they change as you go up in altitude):

Freezing water = 32°F = 0°C

Boiling water = 212°F = 100°C

Thirty-two degrees of Fahrenheit separation

Who came up with the Fahrenheit scale of temperature? You may not be surprised at his last name — Daniel Gabriel Fahrenheit manufactured thermometers in Amsterdam in the 18th century. For the zero point of his scale, he used the temperature of a bath of ice in a solution of salt (which was a common way of reaching low temperatures in an 18th-century laboratory). He chose 32°F as the temperature of melting ice (going up) and freezing water (going down). Using this scale of temperature, water boils at 212°F.

Crossing Fahrenheit and Celsius

The Fahrenheit and Celsius scales actually cross at one point, where the temperature in Fahrenheit and Celsius is the same. What's that temperature? If you call it t, you know that $t = (9 / 5) t + 32$. Solving for t gives you $t = -40$. So, $-40°C = -40°F$, a rather peculiar fact.

If you do the math, you find 180°F between the points of freezing and boiling in the Fahrenheit system and 100°C in the Celsius system, so the conversion ratio is 180 / 100 = 18 / 10 = 9 / 5. And don't forget that the measurements are also offset by 32 degrees (the 0-degrees point of the Celsius scale corresponds to the 32-degrees point of the Fahrenheit scale). Putting these ideas together lets you convert from Celsius to Fahrenheit or from Fahrenheit to Celsius pretty easily; just remember these equations:

$$C = (5 / 9)(F - 32)$$
$$F = (9 / 5) C + 32$$

For example, the blood temperature of a healthy human being is 98.6°F. What does this equal in Celsius? Just plug in the numbers:

$$C = (5 / 9)(F - 32) = (5 / 9)(98.6 - 32) = 37.0°C$$

When the thermometer says Kelvin

William Thompson created a third temperature system, one now in common use in physics, in the 19th century — the Kelvin system (he later became Lord Kelvin). The Kelvin system has become so central to physics that the Fahrenheit and Celsius systems are defined in terms of the Kelvin system — a system based on the concept of absolute zero.

Analyzing absolute zero

Temperature is really a measure of molecular movement — how fast and how much the molecules of whatever object you're measuring are moving. The molecules move more and more slowly as the temperature lowers. At *absolute zero,* the molecules stop, which means that you can't cool them any more. No refrigeration system in the world — or in the entire universe — can go any lower.

The Kelvin system is based on absolute zero as its zero point, which makes sense when you think about it. What's a little odd is that you don't measure temperature in this scale in degrees; you measure it in *kelvins* (seems like Lord Kelvin wanted to make sure you never forget his name). A temperature of 100° in the Celsius system is 100°C, but a temperature of 100 is 100 kelvins in the Kelvin scale. This system has become so widely adopted that the official meters-kilograms-seconds (MKS) unit of temperature is the kelvin (in practice, you see °C used more often in introductory physics).

Making Kelvin conversions

Each kelvin is the same size as a Celsius degree, which makes converting between Celsius degrees and kelvins easy. On the Celsius scale, absolute zero is –273.15°C. This temperature corresponds to 0 kelvins, which you also write as 0K (not, please note, 0°K).

So, to convert between the Celsius and Kelvin scales, all you have to use is the following formula:

$$K = C + 273.15$$
$$C = K - 273.15$$

And to convert from kelvins to Fahrenheit, you can use this formula:

$$F = (9 / 5)(K - 273.15) + 32 = (9 / 5)K - 459.67$$

What temperature does water boil at in kelvins? Well, pure water boils at 100°C, so

$$K = C + 273.15 = 100 + 273.15 = 373.15K$$

Water boils at 373.15K.

Helium turns to liquid at 4.2K; what's that in °C? Use the formula:

$$C = K - 273.15 = 4.2 - 273.15 = -268.95°C$$

Helium liquefies at –268.95°C. Pretty chilly.

The Heat Is On: Linear Expansion

When you talk about the expansion of a solid in any one dimension under the influence of heat, you're talking about *linear expansion*. For example, some screw-top jars can be tough to open, which is maddening when you really want some peanut butter or pickles. Maybe you remember your mom running the lids of jars under hot water as a kid. She did this because the hot water makes the lid expand, which usually makes the job of turning it much easier. Physics takes an interest in this phenomenon, and artists draw figures of it, like the one you see in Figure 13-1.

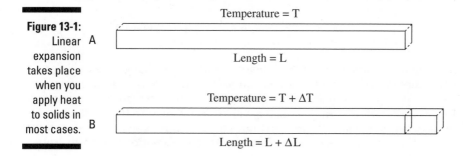

Figure 13-1:
Linear
expansion
takes place
when you
apply heat
to solids in
most cases.

If you raise the temperature of an object a small amount:

$$T_f = T_o + \Delta T$$

where T_f represents the final temperature, T_o represents the original temperature, and ΔT represents the change in temperature; linear expansion results in an expansion in any linear direction of

$$L_f = L_o + \Delta L$$

where L_f represents the final length of the solid, L_o represents its original length, and ΔL represents the change in length. If the temperature goes down a small amount:

$$T_f = T_o - \Delta T$$

you get a linear contraction rather than an expansion:

$$L_f = L_o - \Delta L$$

In other words, the change in length, ΔL, is proportional to the change in temperature, ΔT.

This proportion isn't true for all solids, however — just for most. Some solids contract when you heat them. For example, ice actually contracts as you raise its temperature from 0°C to 4°C as its molecules rearrange themselves from the crystalline structure of ice.

Deconstructing linear expansion

On a molecular level, linear expansion happens because when you heat objects up, the molecules bounce around faster, which leads to a physical expansion. When you heat up a solid, it expands by a few percent, and that percentage is proportional to the change in temperature, so you can write:

$\Delta L / L_o$ (the fraction the solid expands) is proportional to ΔT (the change in temperature)

The constant of proportionality depends on what material you're working with; therefore, in practice, the constant is something you measure, much like the coefficient of friction (see Chapter 6). And like the coefficient of friction, the constant of proportionality is also a coefficient — the coefficient of linear expansion, which is given by the symbol α (not to be confused with the symbol for angular acceleration). You can write this relationship as an equation this way:

$$\Delta L / L_o = \alpha \Delta T$$

You usually write the equation in this form:

$$\Delta L = \alpha L_o \Delta T$$

You usually measure α, the coefficient of linear expansion, in $1/°C$ (that is, in $°C^{-1}$).

Workin' on the railroad: A linear expansion example

Say that you're called in to check out a new railroad. Given the power generated by the engine, the steepness of the grade, and the mass of the load, you calculate that the engine should be able to make the climb at 1.0 meter per second, no problem.

"Huh," says the chief designer. "I could've told you everything was okay. No need to bring in some high-priced physics specialist." As his crew laughs, you look closer at the 10.0-meter-long rails, noticing that they're only 1 millimeter apart at the ends. "How much hotter does it get around these parts during the summer?"

"Hotter?" the designer guffaws. "You afraid the rails will *melt?*"

Everyone snickers at your ignorance as you check your almanac, which tells you that you can expect the rails to get 50°C hotter during a normal summer. The coefficient of linear expansion for the steel that the rails are made from is approximately 1.2×10^{-5} °C^{-1}. (Don't worry; you'll always be given these coefficients when you need them to solve a problem. Just in case, here's a useful Web site that lists many of the coefficients: www.engineeringtoolbox.com/linear-expansion-coefficients-24_95.html.) So, how much will the typical rail expand during the hot part of summer? You know that

$$\Delta L = \alpha L_o \Delta T$$

Plugging in the numbers gives you

$$\Delta L = \alpha L_o \Delta T = (1.2 \times 10^{-5})(10.0)(50) = 6.0 \times 10^{-3} \text{ m}$$

In other words, you can expect the rails to expand 6.0×10^{-3} meters in the summer, or about 6 millimeters. However, the rails are only 1 millimeter apart. The railroad company is in trouble.

You look at the chief designer and say, "You and I are about to have a nice long talk about physics."

Plenty of construction projects take linear expansion into account. You often see bridges with "expansion joints" connecting the bridge to the road surface.

The Heat Continues On: Volume Expansion

Linear expansion, as the name indicates, takes place in one dimension, but the world isn't always linear. It comes supplied with three dimensions. If an object undergoes a small temperature change of just a few degrees, you can say that the volume of the solid will change in a way proportionate to the temperature change. As long as the temperature differences involved are small, you get

$\Delta V / V_o$ (the fraction the solid expands) is proportional to ΔT (the change in temperature)

where ΔV represents the change in volume and V_o represents the original volume. With volume expansion, the constant involved is called the *coefficient of volume expansion*. This constant is given by the symbol β, and like α, it's measured in °C^{-1}. Using β, here's how you can express the equation for volume expansion:

$$\Delta V / V_o = \beta \Delta T$$

When you multiply both sides by V_o, you get

$$\Delta V = \beta V_o \Delta T$$

You've created the analog (or equivalent) of the equation $\Delta L = \alpha L_o \Delta T$ for linear expansion (see the previous section).

If the distances involved are small and the temperature changes are small, you usually find that $\beta = 3\alpha$. This makes sense, because you go from one dimension to three. For example, α for steel is 1.2×10^{-5} $°C^{-1}$, and $\beta = 3.6 \times 10^{-5}$ $°C^{-1}$. This may save you some time in the future.

Say, for example, you're at the gasoline refinery when you notice that workers are filling all the 5,000-gallon tanker trucks to the very brim before leaving the refinery on a hot summer day.

"Uh oh," you think as you get your calculator out. For gasoline, $\beta = 9.5 \times 10^{-4}$ $°C^{-1}$, and you figure that it's $10°C$ warmer in the sunshine than in the building, so

$$\Delta V = \beta V_o \Delta T = (9.5 \times 10^{-4})(5{,}000)(10) = 47.5 \text{ gallons}$$

Not good news for the refinery — those 5,000-gallon tankers of gasoline that are filled to the brim have to carry 5,047.5 gallons of gasoline after they go out in the sunshine. The gas tanks may also expand, but the β of steel is much less than the β of gasoline. Should you tell the refinery workers? Or should you ask for a bigger fee first?

Going with the Flow (of Heat)

What, really, is heat? When you touch a hot object, heat flows from the object to you, and your nerves record that fact. When you touch a cold object, heat flows from you to that object, and again, your nerves keep track of what's happening. Your nerves record why objects feel hot or cold — because heat flows from them to you or from you to them.

So, what is heat in physics terms? *Heat* is energy that flows from objects of higher temperatures to objects of lower temperatures. The unit of this energy in the meters-kilograms-seconds (MKS) system is the Joule (J).

Bundling into a blanket

Why does a blanket keep you warm? Aren't you transferring your heat through the blanket into the outside air?

You certainly are, but when you warm up the mass of the blanket and the air trapped in it, that warm material transfers heat in all directions, not just to the outside air. Some of the heat comes back to you. The thicker the blanket, the more heat that has a chance to transfer back to you as it travels from layer to layer through the blanket.

Where does this energy come from? Take a look at the molecular picture — heat is a measure of the energy stored in the internal molecular motion of an object. Heat stays in an object until it can leak out to the environment.

Different materials can hold different amounts of heat — if you warm up a potato, it can hold its heat longer (as your tongue can testify) than a lighter material like cotton candy. The measure of how much heat an object can hold is called its *specific heat capacity.*

Physicists like to measure everything, so it shouldn't be surprising that you bust out your thermometer when you see someone making a cup of coffee. You measure exactly 1.0 kg of coffee in the pot, and then you get down to the real measurements. You find out that it takes 4,186 J of heat energy to raise the temperature of the coffee by 1°C, but it takes only 840 J to raise 1.0 kg of glass by 1°C. Where's that energy going? It goes into the object being heated, which stores it as internal energy until it leaks out again. *Note:* If it takes 4,186 J to raise 1.0 kg of coffee by 1°C, it takes double that, 8,372 J, to raise 2.0 kg of coffee by 1°C.

You can relate the amount of heat, Q (the common letter to use for heat; yep, it makes no sense), it takes to raise the temperature of an object to the change in temperature and the amount of mass involved:

$$Q = cm\Delta T$$

where Q is the amount of heat energy involved (measured in Joules if you're using the MKS system), m is the amount of mass, ΔT is the change in temperature, and c is a constant called the *specific heat capacity,* which is measured in J/(kg-°C) in the MKS system.

One *calorie* is defined as the amount of heat needed to heat 1.0 gram of water 1.0°C, so 1 calorie = 4,186 J. Nutritionists use the food-energy term Calorie (capital C) to stand for 1,000 calories — 1.0 kcal — so 1.0 Calorie = 4,186 J. Physicists use another unit of measurement as well — the British Thermal Unit (Btu). One Btu is the amount of heat needed to raise 1 pound of water 1.0°F. To convert, you can use the relation that 1 Btu = 1,055 J.

Now that the 45 g of coffee in your cup is cold, you call over your host. "It's 45°C," you say, "but I like it at 65°C." The host gets up to pour some more.

"Just a minute," you say. "The coffee in the pot is 95°C. Wait until I calculate exactly how much you need to pour."

The following represents the heat lost by the new mass of coffee, m_1:

$$\Delta Q_1 = cm_1 (T_f - T_{1o})$$

And here's the heat gained by the existing coffee, mass m_2:

$$\Delta Q_2 = cm_2 (T_f - T_{2o})$$

Assuming you have a super-insulating coffee mug, the heat lost by the new coffee is the heat that the existing coffee will gain, so

$$\Delta Q_1 = -\Delta Q_2$$

which means that

$$-cm_1 (T_f - T_{1o}) = cm_2 (T_f - T_{2o})$$

Dividing both sides by the specific heat capacity and plugging in the numbers gives you

$$-m_1 (65 - 95) = -m_1 (-30) = m_1(30)$$

$$m_2 (T_f - T_{2o}) = (0.045)(65 - 45) = 0.9$$

$$m_1(30) = 0.9$$

Dividing both sides by 30 gives you

$$[m_1(30)] / 30 = 0.9 / 30 = m_1 = 0.03 \text{ kg} = 30 \text{ g}$$

Satisfied, you put away your calculator and say, "Give me exactly 30.0 grams of that coffee."

Changing Phases: When Temperatures Don't Change

The equation of the moment is $Q = cm\Delta T$ — used to find the amount of heat energy with respect to the change in temperature — but you also come across cases where adding or removing heat doesn't change the temperature. For example, imagine you're calmly drinking your lemonade at an outdoor garden party. You grab some ice to cool your lemonade, and the mixture in your glass is now half ice, half lemonade (which you can assume has the same specific heat as water; see the previous section), with a temperature of exactly 0°C.

As you hold the glass and watch the action, the ice begins to melt — but the contents of the glass don't change temperature. Why? The heat going into the glass is melting the ice, not warming the mixture up. So, does this make the equation for heat energy useless? Not at all. It just means that the equation doesn't apply for a phase change.

Phase changes occur when materials change state, from liquid to solid (as when water freezes), solid to liquid (as when rocks melt into lava), liquid to gas (as when you boil tea), and so on. When the material in question changes to a new state — liquid, solid, or gas (you can also factor in a fourth state: plasma, a superheated gas-like state) — some heat goes into or comes out of the process.

You can even have solids that turn directly into gas, as when you have a block of dry ice, frozen carbon dioxide. As the dry ice gets warmer, it turns into carbon dioxide gas. This process is called *sublimation*.

Breaking the ice with phase changes

Imagine that someone has taken a bag of ice and thoughtlessly put it on the stove. Before it hit the stove, the ice was at a temperature below freezing (–5°C), but being on the stove is about to change that. You can see the change taking place in graph form in Figure 13-2.

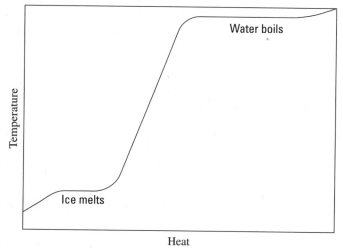

Water boils

Temperature

Ice melts

Heat

Figure 13-2:
Phase changes of water.

As long as no phase change takes place, the equation $Q = cm\Delta T$ holds good [the specific heat capacity of ice is around 2.0×10^3 J/(kg-°C)], which means that the temperature of the ice will increase linearly as you add more heat to it, as you see in the graph in Figure 13-2.

However, when the ice reaches 0°C, you start to get nervous. The ice is getting too warm to hold its solid state, and it begins to melt, undergoing a phase change. On a molecular level, however, it takes energy to make an object change state. For example, when you melt ice, it takes energy to break up the crystalline ice structure. The energy needed to melt the ice is supplied as heat. This is why the graph in Figure 13-2 levels off in the middle — the ice is melting. It takes heat to make the ice change phase to water, so even though the stove adds heat, the temperature of the ice doesn't change as it melts.

As you watch the bag of ice on the stove, however, you note that all the ice eventually melts into water. Because the stove is still adding heat, the water begins to heat up, which you see in Figure 13-2. The stove adds more and more heat to the water, and in time, the water starts to bubble. "Aha," you think. "Another phase change." And you're right: The water is boiling and becoming steam. The bag holding the ice seems pretty resilient, however, and expands while the water turns to steam; but it doesn't break.

You measure the temperature of the water. Fascinating — although the water boils, turning into steam, the temperature doesn't change, as shown in Figure 13-2. Once again, you need to add heat to incite a phase change — this time from water to steam. You can see in Figure 13-2 that as you add heat, the water boils, but the temperature of that water doesn't change.

What's going to happen next, as the bag swells to an enormous volume? You never get to find out, because the bag finally explodes. You pick up a few shreds of the bag and examine them closely. How can you account for the heat that's needed to change the state of an object? How can you add something to the equation for heat energy to take into account phase changes? That's where the idea of latent heat comes in.

Understanding latent heat

Latent heat is the heat per kilogram that you have to add or remove to make an object change its state; in other words, latent heat is the heat needed to make a phase change happen. Its units are J/kg in the meters-kilograms-seconds (MKS) system.

Physics recognizes three types of latent heat, corresponding to the changes of phase between solid, liquid, and gas:

- ✔ **The latent heat of fusion, L_f:** The heat per kilogram needed to make the change between the solid to liquid phases, as when water turns to ice.

- ✔ **The latent heat of vaporization, L_v:** The heat per kilogram needed to make the change between the liquid to gas phases, as when water boils.

- ✔ **The latent heat of sublimation, L_s:** The heat per kilogram needed to make the change between the solid to gas phases, as when dry ice evaporates.

For example, the latent heat of fusion of water, L_f, is 3.35×10^5 J/kg, and the latent heat of vaporization of water, L_v, is 2.26×10^6 J/kg. In other words, it takes 3.35×10^5 Joules to melt 1 kilogram of ice at 0°C (just to melt it, not to change its temperature). And it takes 2.26×10^6 Joules to boil 1 kilogram of water into steam. These are the energies involved in making water change its phase.

Here is a Web site that lists some common numbers for fusion and vaporization: www.physchem.co.za/Heat/Latent.htm#fusion

Chapter 14

Here, Take My Coat: Heat Transfer in Solids and Gases

*Y*ou witness the transfer of heat and the chemical changes that result every day. You cook some pasta, and you see currents of water cycling the noodles in the pan. You pick up the pan without a hand towel, and you burn your hand. You look to the sky on a summer day, and you feel your face warming up. You give your coat to your dates, and you watch their feelings for you warm up (through radiation, of course!).

In this chapter, I discuss the three primary ways by which heat is distributed. I start by discussing the ways in which heat can be transferred, and then I discuss how heat affects gases — which is where you find out all about moles and how to predict how many molecules are present in certain liquids or solids. You also find out how heat moves around, how many molecules you're dealing with in certain situations, and more. After you read this chapter, you can startle people at parties by telling them how many molecules of water they have in their glasses!

Boiling Water: Convection

Convection is one of the primary means by which heat is transferred from point to point. It happens when a substance like air or water is heated; some of it becomes less dense than what's left, and the heated part rises. Take a look at the pot of water heating up in Figure 14-1. How does the heat move around in the water? As the water near the heating element heats up, it expands and becomes less dense. The less-dense water rises toward the top, which starts a current. You can see convection at work if you add a few noodles to the pot and watch how they circulate. Here, the heat transfer relies on a medium that expands and moves around — water.

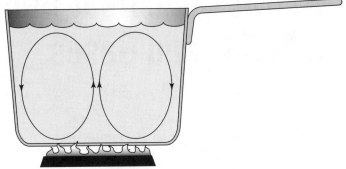

However, convection doesn't need water; air and many other substances do the trick. Maybe you've seen birds riding *thermals* (columns of rising hot air) while circling and rising higher in the air — *convective motion* of the air that was heated near the Earth's surface created the thermals.

Many ovens work by using convection; you may hear the term convection oven as opposed to, say, microwave oven (a microwave oven heats by using microwaves to briefly polarize the molecules in food and make them oscillate; so, believe it or not, the heating in a microwave actually comes from friction). The air inside a convection oven gets hot, which makes air currents circulate.

You may have heard the maxim "heat rises," which is all about convection. Hot air expands and becomes less dense than the cool air around it, which makes it rise. So, if you have a two- or three-story house with an open stairway, all your carefully heated air concentrates happily on the top floor during the winter. The heat from radiators in a room also rises by convection, circulating the air around to some extent. To increase the heat movement, people sometimes use ceiling fans. You probably see convection on a daily basis, and now you can recognize it in action; however, there are other heating events taking place that you still haven't discovered. Read on!

Too Hot to Handle: Conduction

Conduction transfers heat through material directly, without the need of currents to sustain the transfer (like convection; see the preceding section). Take a look at the metal pot in Figure 14-2 and its metal handle; the pot has been boiling for 15 minutes. Would you want to lift it off the fire by grabbing the handle without an oven mitt? Probably not, unless you're a thrill-seeker.

The metal handle is hot, but why? Not because of convection, because no heat-driven currents of mass are at work here. The handle is hot because of conduction.

On the molecular level, the molecules near the heat source are heated and begin moving faster. They bounce off the molecules near them and cause them to move faster (as anyone who's played billiards can attest). That increased bouncing is what heats a substance up.

Figure 14-2: Conduction heats the pot that holds the boiling water.

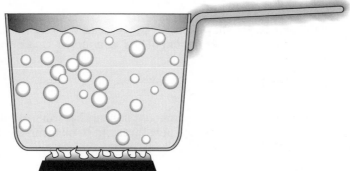

Some materials, like most metals, conduct heat better than others, such as porcelain, wood, or glass. The way substances conduct heat depends a great degree on their molecular structures, so different substances react differently.

Examining the properties that affect conduction to find the conduction equation

You have to take different properties of objects into account when you want to examine the conduction that takes place. If you have a bar of steel, for example, you have to take into account the bar's area and length, along with the temperature at different parts of the bar. Take a look at Figure 14-3, where a bar of steel is being heated on one end and the heat is traveling by conduction toward the other side. Can you find out how fast the heat travels? No problem.

You can go a couple different routes in this situation. For example, as you may expect, the greater the difference in temperature between the two ends of the bar, the greater the amount of heat that's transferred. It turns out that the amount of heat transferred, Q, is proportional to the difference in temperature, ΔT ($\propto$ means "is proportional to"):

$$Q \propto \Delta T$$

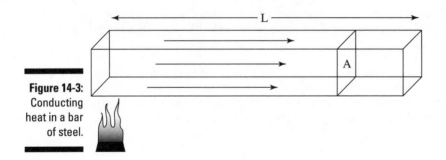

Figure 14-3:
Conducting
heat in a bar
of steel.

On the other hand, as you may expect, a bar twice as wide conducts twice the amount of heat. In general, the amount of heat conducted, Q, is proportional to the cross-sectional area, A, like this:

$$Q \propto A$$

And the longer the bar, the less the amount of heat that will make it all the way through; in fact, the conducted heat turns out to be inversely proportional to the length of the bar, L:

$$Q \propto 1 / L$$

Finally, the amount of heat transferred, Q, depends on the amount of time that passes, t — twice the time, twice the heat. Here's how you express this mathematically:

$$Q \propto t$$

When you put it all together and nominate k as the constant that's yet to be determined, the heat transfer by conduction through a material is given by this equation:

$$Q = (kA\Delta Tt) / L$$

This represents the amount of heat transferred by conduction in a given amount of time, t, down a length L, where the cross-sectional area is A. Here, k is the material's *thermal conductivity,* measured in Joules/(s-m-°C).

REMEMBER

Different materials (glass, steel, copper, bubble gum) conduct heat at different rates, so the thermal conductivity constant depends on the material in question. Lucky for you, physicists have measured the constants for various materials already. Check out some of the calculations in Table 14-1.

Table 14-1	Thermal Conductivities for Various Materials
Material	*Thermal Conductivity*
Glass	0.80 J/(s-m-°C)
Steel	14.0 J/(s-m-°C)
Copper	390 J/(s-m-°C)
Styrofoam	0.01 J/(s-m-°C)
Brass	110 J/(s-m-°C)
Silver	420 J/(s-m-°C)

Applying the heat-transferred-by-conduction equation

The thermal conductivity of the steel part of a pot handle is 14.0 Joules/(s-m-°C) (see Table 14-1). Take a look at Figure 14-3. If that handle is 15 cm long, with a cross sectional area of 2.0 cm^2 (2.0×10^{-4} m^2), and the fire at one end is 600°C, how much heat would be pumped into your hand if you grabbed it? The equation for heat transfer by conduction is

$$Q = (kA\Delta Tt) / L$$

If you assume that the end of the cool end of the handle starts at about room temperature, 25°C, you get

$$Q = (kA\Delta Tt) / L = [(14)(2.0 \times 10^{-4})(600 - 25)t] / 0.15 = (10.7J)t$$

Becoming a cooler engineer

Based on the thermal conducting or insulating abilities of different materials, you can calculate how to keep objects cold or hot, from drinks to hot dogs to racks of meat; for example, imagine you've been asked to design a frozen food storage locker, and your first thought is to make it out of copper, which has a thermal conductivity of 390J/(s-m-°C). However, copper would give disappointing results, because it has a very high thermal conductivity, which would let heat move in or out of the locker easily. Better check the thermal conductivities of different objects. Styrofoam has a thermal conductivity of only 0.01 J/(s-m-°C), which is a much better choice; if ice lasts a time *t* in the copper locker, it will last 390 / .01 = 39,000 times longer if you build the cooler with Styrofoam instead. (That's just an estimate, of course; the actual answer depends on the thermal conductivity of the contents of the locker as well).

You find that 10.7 Joules (the unit of measure in question) of heat are being transferred to the end of the handle each second. And as the seconds go by, the Joules of heat will add up, making the handle hotter and hotter.

Emitting and Absorbing Light: Radiation

You get out of the shower soaking wet in the dead of winter, and you're toasty warm. Why? Because of a little physics, of course. In particular, the heat lamp you have in your bathroom, which you can see in Figure 14-4. The heat lamp beams out heat to you and keeps you warm through radiation.

Figure 14-4:
A light bulb radiates heat into its environment.

Radiation is light that can transfer heat. Heat energy transferred through radiation is as familiar as the light of day; in fact, it *is* the light of day. The sun is a huge thermal reactor about 93 million miles away in space, and neither conduction nor convection (see the previous two sections in this chapter) would produce any of the energy that arrives to Earth through the vacuum of space. The sun's energy gets to the Earth through radiation, which you can confirm on a sunny day just by standing outside and letting the sun's rays warm your face. The only way you can actually see radiation, however, is through the sunburn you later have to manage.

You can't see radiation, but it's there

Every object around you is continually radiating, unless it's at absolute zero temperature, which is a little unlikely because you can't physically get to a temperature of absolute zero. A scoop of ice cream, for example, radiates. Electromagnetic radiation comes from accelerating and decelerating electric charges, and on a molecular level, that's what happens as objects warm up — their atoms move around faster and faster and bounce off other atoms hard.

Even you radiate all the time, but that light isn't usually visible because it's in the infrared part of the spectrum. However, that light is visible to infrared scopes, as you've seen in the movies or television. You radiate heat in all directions all the time, and everything in your environment radiates heat back to you. When you have the same temperature as your surroundings, you radiate as fast and as much to your environment as it does to you.

If your environment didn't radiate heat back to you, you'd freeze, which is why space is considered so "cold." There's nothing cold to touch in space, and heat isn't lost through conduction or convection. All that happens is that the environment doesn't radiate back at you, which means that the heat you radiate away is lost. And you can freeze very fast from the lost heat.

When an object heats up to about 1,000 kelvin, it starts to glow red (which may explain why, even though you're radiating, you don't glow red in the visible light spectrum). As the object gets hotter, its radiation moves up in the spectrum through orange, yellow, and so on up to white hot at somewhere around 1,700K.

Radiant heaters with coils that glow red rely on radiation to transfer heat. Convection takes place as air gets heated, rises, and spreads around the room (and conduction can occur if you touch the heater on a hot spot by mistake — not the most desirable of heat transfers). But heat transfer takes place mostly through radiation. Many houses have heating wires buried in walls, ceilings, or floors, which is called *radiant heating* — you don't see anything (which, architecturally, is the idea), but your face will feel warm as you face the radiant heat source.

Humans understand heat radiation and absorption in the environment intuitively. For example, on a hot day, you may avoid wearing a black t-shirt, because you know it will make you hotter. Why? Because it absorbs light from the environment while reflecting less of it back than a white t-shirt. The white t-shirt keeps you cooler, because it reflects more radiant heat back to the environment. Which would you rather get into on a hot day: a car with black leather upholstery or one with white?

Reflecting on cold metal

Why do metals feel cold to the touch? When you pick up a metal object (even one you don't find in the refrigerator, such as a can of soda), it feels cold to the touch. You probably take this for granted in your everyday life. But why is that?

The answer lies in radiant heat. Metals are shiny, so they can reflect back the radiant heat hitting them better than other objects in a room, even white ones. So, if a spoon reflects back 90 percent of the radiant heat hitting it, and a book reflects back just 10 percent, which feels cooler to the touch? Metal, indeed.

Radiation and blackbodies

Some objects absorb more of the light that hits them than others. The objects that absorb *all* the radiant heat that strikes them have a name — they're called *blackbodies.* A blackbody absorbs 100 percent of the radiant heat striking it, and if it's in equilibrium with its surroundings, it emits all the radiant heat as well.

Most objects fall between mirrors, which reflect all light, and blackbodies, which absorb all light. The middle-of-the-road objects absorb some of the light striking them and emit it back into their surroundings. Shiny objects are shiny because they reflect most of the light, which means they don't have to emit as much heat radiantly into the room as other objects. Dark objects appear dark because they don't reflect much light, which means they have to emit more as radiant heat (usually lower down in the spectrum, where light is infrared and can't be seen).

You can find plenty of physics when it comes to blackbodies, starting with the following question: How much heat does a blackbody emit when it's at a certain temperature? The amount of heat radiated is proportional to the time you allow — twice as long, twice as much heat radiated, for example. So, you can write the equation, where t is time, as follows:

$$Q \propto t$$

And as you may expect, the amount of heat radiated is proportional to the total area doing the radiating. So, you can also write the equation as follows, where A is the area doing the radiating:

$$Q \propto At$$

Temperature, T, has to be in the equation somewhere — the hotter an object, the more heat radiated. Experimentally, it turns out that the amount of heat radiated is proportional to T to the fourth power, T^4. So, now you have

$$Q \propto AtT^4$$

To make the equation final, you need to add a constant, which is measured experimentally. To find the heat emitted by a blackbody, you use the Stefan-Boltzmann constant, σ, which goes in like this:

$$Q = \sigma AtT^4$$

The value of σ is 5.67×10^{-8} J(s-m^2-K^4). Note, however, that this constant works only for blackbodies that are perfect emitters. Most objects aren't perfect emitters, so you have to add another constant most of the time — one that depends on the substance you're working with. The constant is called *emissivity, e*. So, this equation — the *Stefan-Boltzmann law of radiation* — becomes

$$Q = e\sigma AtT^4$$

where *e* is an object's emissivity, σ is the Stefan-Boltzmann constant (5.67×10^{-8} J[s-m^2-K^4]), A is the radiating area, *t* is time, and T is the temperature in kelvins.

For example, a person's emissivity is about 0.8. At a body temperature of 37°C, how much heat does a person radiate each second? First, you have to factor in how much area does the radiating. If you consider a person mathematically as a 1.6-meter-high cylinder of radius 0.1m, the total surface area would be

$$A = (\text{area of a cylinder}) = 2\pi rh + 2\pi r^2 = 2\pi\,(0.1)(1.6) + 2\pi(0.1)^2 = 1.07 \text{ m}^2$$

where *r* is the radius and *h* is height. To find the total heat radiated by a person, plug the numbers into the Stefan-Boltzmann law of radiation equation:

$$Q = e\sigma AtT^4 = (0.8)(5.67 \times 10^{-8})(1.07)(37 + 273.15)^4 t = 449t$$

You get a value of 449 Joules per second, or 449 watts. That may seem high, because skin temperature isn't the same as internal body temperature, but it's in the ballpark.

Crunching Avogadro's Number

Much of the info in this chapter deals with the effects of heat on solids and liquids (as in pools of liquid metal, for example), but you can examine plenty of physics when you heat up gases, too. It all starts by knowing how many molecules of a substance you're dealing with, which you can figure out by using a little physics. For example, consider the scenario of a person who finds a large diamond and brings it to you for examination.

"How many atoms are in my diamond?"

"That depends," you say, "on how many moles you have."

The person, offended, says, "I *beg* your pardon!"

A *mole* is the number of atoms in 12.0 grams of carbon isotope 12. Carbon isotope 12 — also called carbon-12, or just carbon 12 — is the most common version of carbon, although some carbon atoms have a few more neutrons in them — carbon 13, actually — so the average works out to about 12.011. You have to know how many moles of matter you have before you can know how many atoms you have, and an atom-by-atom knowledge is essential when you start working with heat and gases.

The number of atoms has been measured as 6.022×10^{23}, which is called Avogadro's Number, N_A. So, now you know how many atoms are in 12.0 grams of carbon 12. Do you find the same number of atoms in, say, 12.0 grams of sulphur? Nope. Each sulpher atom has a different mass from each carbon atom, so even if you have the same number of grams, you have a different number of atoms.

How much more mass does sulphur have than carbon 12? If you check a periodic table of elements hanging on the wall in a physics lab, you'll find that the atomic mass of sulphur (usually the number appearing right under the element's symbol, like S for sulphur) is 32.06. But 32.06 what? It's 32.06 *atomic units, u,* where each atomic unit is $\frac{1}{12}$ of the mass of a carbon 12 atom. If a mole of carbon 12 has a mass of 12.0 grams, therefore, and the mass of your average sulphur atom is bigger than the mass of a carbon 12 atom by the following ratio:

Sulphur mass = 32.06

Carbon 12 mass = 12 u

a mole of sulphur atoms must have this mass:

$$(32.06 / 12 \text{ u})(12.0 \text{ g}) = 32.06 \text{ g}$$

How convenient! Knowing that a mole of an element has the same mass in grams as its atomic mass in atomic units proves helpful in your calculations. You can read the atomic mass of any element in atomic units off any periodic table. You can find that a mole of silicon has a mass of 28.09 grams, a mole of sodium has a mass of 22.99 grams, and so on. And each of those moles contains 6.022×10^{23} atoms.

Now you can determine the number of atoms in a diamond, which is solid carbon (atomic mass = 12.01 u). A mole is 12.01 g of diamond, so when you find out how many moles you have, you multiply that times 6.022×10^{23} atoms.

Not every object is made up of a single kind of atom. When was the last time someone handed you a kilogram of 100 percent-pure sulphur on the street? Most materials are composites, such as water, which is made up of two hydrogen atoms for every one oxygen atom (H_2O). In cases like these, instead of the atomic mass, you look for the *molecular mass* (when atoms combine, you have molecules), which is also measured in atomic mass units. For example, the molecular mass of water is 18.0153 u, so 1 mole of water molecules has a mass of 18.0153 g.

Forging the Ideal Gas Law

When you start working on an atom-by-atom and molecule-by-molecule level, you begin working with gases from a physics point of view. For example, how do moles factor into predicting what's going to happen when you heat gases? It turns out that you can relate the temperature, pressure, volume, and number of moles together for a gas. The relation doesn't always hold completely true, but it always does for *ideal gases*. Ideal gases are very light, like helium, and they're the frictionless ramps of thermodynamics.

It's an experimental fact that if you keep the volume constant and heat a gas, the pressure will go up linearly, as you see in Figure 14-5. In other words, at a constant volume, where T is the temperature measured in kelvins and P is the pressure,

$$P \propto T$$

You can let the volume vary as well, and if you do, you'll find that the pressure is inversely proportional to the volume — if the volume of a gas doubles, its pressure will go down by a factor of two:

$$P \propto T / V$$

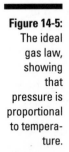

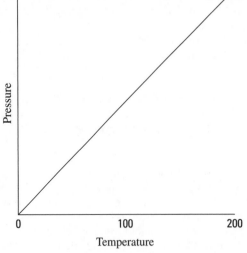

On the other hand, when the volume and temperature of an ideal gas are constant, the pressure is proportional to the number of moles of gas you have — twice the amount of gas, twice the pressure. If the number of moles is n,

$$P \propto nT / V$$

Adding a constant, R — the *universal gas constant,* which has a value of 8.31 Joules(mole-K) — gives you the *ideal gas law,* which relates pressure, volume, number of moles, and temperature:

$$PV = nRT$$

Ideal gases are those gases for which the ideal gas law holds. By using the ideal gas law, you can predict the pressure of an ideal gas, given how much you have of it, its temperature, and the volume you've enclosed it in.

You can also express the ideal gas law a little differently by using Avogadro's Number, N_A (see the previous section), and the total number of molecules, N:

$$PV = nRT = (N / N_A) RT$$

The constant R/N_A is also called Boltzmann's constant, k, and has a value of 1.38×10^{-23} J/K. Using this constant, the ideal gas law becomes

$$PV = NkT$$

Gas pressure: An ideal gas law example

Say that you're measuring a volume of 1 cubic meter filled with 600 moles of helium (which is very close to an ideal gas) at room temperature, 27°C. What would be the pressure of the gas? Using the following form of the ideal gas law:

$$PV = nRT$$

you can put P on one side by dividing by V. Now plug in the numbers:

$$P = nRT / V = [(600.0)(8.31)(273.15 + 27)] / 1.0 = 1.50 \times 10^6 \text{ N/m}^2$$

The pressure on all the walls of the container is 1.50×10^6 N/m^2. Notice the units of pressure here — N/m^2. The unit is used so commonly that it has its own name in the MKS system: Pascals, or Pa. One Pa = 1.45×10^{-4} pounds per inch^2. Atmospheric pressure is 1.013×10^5 Pa, which is 14.70 lb/in^2. The pressure of 1 atmosphere is also given in *torr* on occasion, and 1.0 atmosphere = 760 torr. In this example, you have a pressure of 1.50×10^6 Pa, which is about 15 atmospheres.

You may come across a special set of conditions sometimes used when talking about gases — Standard Temperature and Pressure, also called STP. The STP pressure is 1 atmosphere, 1.013×10^5 Pa, and the STP temperature is 0°C. You can use the ideal gas law to calculate that at STP, 1.0 mole of an ideal gas occupies 22.4 liters of volume (1.0 liter is 1×10^{-3} m^3).

Boyle's Law and Charles' Law: Alternative expressions of the ideal gas law

You can often express the ideal gas law in different ways. For example, you can express the relationship between the pressure and volume of an ideal gas before and after one of those quantities changes at a constant temperature like this:

$$P_f V_f = P_o V_o$$

This equation, called *Boyle's Law,* says that, all factors being the same, the product PV will be conserved. Similarly, at constant pressure, you can say that

$$V_f / T_f = V_o / T_o$$

This equation, called *Charles' Law,* says that the ratio of V/T will be conserved for an ideal gas, all factors being the same.

Measuring water and air pressure

Want a way to display your pressure expertise in the real world? Well, you can determine the pressure under fluids like water with a simple equation:

$$P = \rho g h$$

Here, ρ is the density of the fluid, g is the normal acceleration due to gravity (9.8 meters per second²), and h is the depth of the fluid you're measuring. For water, ρ is about 1,000 kilograms per meter³, so you get

$$P = (1,000.0)(9.8)h = 9,800h$$

For every meter you go under water, you gain about 9,800 Pa of pressure, or about ¹⁄₁₀ of an atmosphere (see the section "Gas Pressure: An ideal gas law example" for this conversion).

This equation works as long as the density of the fluid doesn't vary, which means that it's hard to predict what the pressure due to air will be, because the density of air varies all the way out to space. You often see physics problems that ask how much the air pressure varies as you go up a certain height; in other words, the problems give you a difference in height over which ρ is constant and ask you to calculate the change in pressure.

Tracking Ideal Gas Molecules

You can examine certain properties of molecules of an ideal gas as they zip around. For instance, you can calculate the average kinetic energy of each molecule with a very simple equation:

$$KE_{avg} = \tfrac{3}{2}kT$$

where k is Boltzmann's constant, 1.38×10^{-23} J/K, and T is the temperature. And because you can determine the mass of each molecule if you know what gas you're dealing with (see the section "Crunching Avogadro's Number" earlier in this chapter), you can figure out the molecules' speeds at various temperatures.

Predicting air molecule speed

Imagine you're at a picnic with friends on a beautiful spring day. You have a nice spread — potato salad, sandwiches, and drinks. But after a while, you wonder if your friends forgot to bring any physics. You flop backward and look up at the sky. It all seems a little boring, but then you remind yourself that physics goes on everywhere, all the time, even if you can't see direct results.

For example, you can't see the air molecules whizzing around you, but you can predict their average speeds. You get out your calculator and thermometer. You measure the air temperature at about 28°C, or 301 kelvin (see Chapter 13 for this conversion). You know that for the molecules in the air, you can measure their average kinetic energy with

$$KE_{avg} = \tfrac{3}{2}kT$$

Now plug in the numbers:

$$KE_{avg} = \tfrac{3}{2}kT = \tfrac{3}{2}(1.38 \times 10^{-23})(301) = 6.23 \times 10^{-21} \text{ J}$$

The average molecule has a kinetic energy of 6.23×10^{-21} Joules. On the other hand, the molecules are pretty small — what speed does 6.23×10^{-21} Joules correspond to? Well, you know from Chapter 8 that

$$KE = \tfrac{1}{2}mv^2$$

where m is the mass and v is the velocity, so

$$v = \sqrt{\frac{2KE}{m}}$$

Air is mostly nitrogen, and each nitrogen atom has a mass of about 4.65×10^{-26} kg (you can figure that one out yourself by finding the mass of a mole of nitrogen and dividing by the number of atoms in a mole, N_A). You can plug in the numbers to get

$$v = \sqrt{\frac{2KE}{m}} = 517 \text{ m/s} = 1{,}150 \text{ miles per hour}$$

Yow! What an image; huge numbers of the little guys crashing into you at 1,150 miles per hour every second! Good thing for you the molecules are so small. Imagine if each air molecule weighed a couple of pounds. Big problems.

Calculating kinetic energy in an ideal gas

It's true that atoms have very little mass, but there are many, many of them in gases, and because they all have kinetic energy, the total kinetic energy can pile up pretty fast. How much total kinetic energy can you find in a certain amount of gas? Each molecule has this average kinetic energy:

$$KE_{avg} = \tfrac{3}{2}kT$$

To figure the total kinetic energy, you multiply by the number of molecules you have, which is nN_A, where n is the number of moles:

$KE_{total} = \frac{3}{2}nN_AkT$

N_Ak equals R, the universal gas constant (see the section "Forging the Ideal Gas Law" earlier in this chapter), so this equation becomes

$KE_{total} = \frac{3}{2}nRT$

If you have 600 moles of helium at 27°C, you have this much internal energy wrapped up in thermal movement:

$KE_{total} = \frac{3}{2}nRT = \frac{3}{2}(600.0)(8.31)(273.15 + 27) = 2.24 \times 10^6 \, J$

This converts to a little less than 1 Calorie (the kind of energy unit you find on food wrappers).

Chapter 15

When Heat and Work Collide: The Laws of Thermodynamics

● ●

In This Chapter

▶ Achieving thermal equilibrium

▶ Storing heat and energy under different conditions

▶ Revving up heat engines for efficiency

▶ Dropping close to absolute zero

● ●

*I*f you've ever had an outdoor summer job, you know all about heat and work, a relationship encompassed by the term *thermodynamics*. This chapter brings together those two cherished topics, which I cover in detail in Chapter 8 (work) and Chapter 13 (heat). Thermodynamics has three laws, much like Newton, but it does Newton one better — it also has a zeroth law of thermodynamics. In this chapter, I cover thermal equilibrium through the zeroth law, heat and energy conservation through the first law, heat flow through the second law, and absolute zero with the third law. Time to throw the book at thermodynamics.

Gaining Thermal Equilibrium: The Zeroth Law of Thermodynamics

The laws of thermodynamics start with the zeroth law. You may think that odd, because few other sets of everyday objects start off that way ("Watch out for that zeroth step, it's a doozy . . ."), but you know how physicists love their traditions. *The zeroth law of thermodynamics* says that two objects are in thermal equilibrium if heat can pass between them, but no heat is doing so.

For example, if you and the swimming pool you're in are at the same temperature, no heat is flowing from you to it or from it to you (although the possibility is there). You're in thermal equilibrium. On the other hand, if you jump into the pool in winter, cracking through the ice covering, you won't be in thermal equilibrium with the water. And you don't want to be. (Don't try this physics experiment at home!)

To check on thermal equilibrium, especially in cases of frozen swimming pools that you're about to jump into, you should use a thermometer. You can check the temperature of the pool with the thermometer and then check your temperature. If the two temperatures agree — in other words, if you're in thermal equilibrium with the thermometer, and the thermometer is in thermal equilibrium with the pool — you're in thermal equilibrium with the pool.

What using a thermometer illustrates is that two objects in thermal equilibrium with a third object are also in thermal equilibrium with each other — another way to state the zeroth law.

Among other jobs, the zeroth law sets up the idea of temperature as a thermal equilibrium indicator. The two objects mentioned in the zeroth law are in equilibrium with a third, giving you what you need to set up a scale like the Kelvin scale. From a physics point of view, the zeroth law sets a foundation by saying that no net heat flows between two objects at the same temperature.

Conserving Heat and Energy: The First Law of Thermodynamics

The *first law of thermodynamics* is a statement of energy conservation. It states that energy is conserved. The internal energy in a system, U_o, changes to a final value U_f when heat, Q, is absorbed or released by the system and the system does work, W, on its surroundings (or the surroundings do work on the system), such that

$$U_f - U_o = \Delta U = Q - W$$

In Chapter 8, I present plenty of discussion about conserving energy — mechanical energy, anyway. I show that the total mechanical energy — the sum of the potential and kinetic energy — is conserved. To say that, you have to work with systems where no energy is lost to heat — there could be no friction, for example. All that changes now. Heat is allowed at last (as you can gather from my mention of thermodynamics). Now you can break down the total energy of a system, which includes heat, work, and the internal energy of the system.

These three quantities — heat, work, and internal energy — make up all the energy you need to consider, so their sum is conserved. When you add heat transfer, Q, to a system, and that system doesn't work, the amount of internal energy in the system, which is given by the symbol U, changes by Q. A system can also lose energy by doing work on its surroundings, such as when an engine lifts weight at the end of a cable. When a system does work on its surroundings and gives off no waste heat, its internal energy, U, changes by W. In other words, you're in a position to think in terms of heat as energy, so when you take into account all three quantities — heat, work, and the internal energy — energy is conserved.

The first law of thermodynamics is a powerful one because it ties all the quantities together. If you know two of them, you can find the third.

Calculating conservation

The quantity Q, representing heat transfer, is positive when the system absorbs heat and negative when the system releases heat. The quantity W, representing work, is positive when the system does work on its surroundings and negative when the surroundings do work on the system.

Too many people get confused trying to figure out the positive or negative values of every mathematical quantity. Don't confuse yourself. When working with the first law of conservation, you can work from the idea of energy conservation instead. Say, for example, that a motor does 2,000 Joules of work on its surroundings while releasing 3,000 Joules of heat. By how much does its internal energy change? In this case, you know that the motor does 2,000 J of work on its surroundings, so you know that its internal energy will decrease by 2,000 J. And it also releases 3,000 J of heat while doing its work, so the internal energy of the system decreases by an additional 3,000 J.

Think of values of work and heat flowing out of the system as negative. In the previous example, thinking this way makes the total change of internal energy

$$\Delta U = -2{,}000 - 3{,}000 = -5{,}000 \text{ J}$$

The internal energy of the system decreases by 5,000 J, which makes sense — the system does 2,000 J of work on the surroundings and releases 3,000 J of heat. On the other hand, what if the system *absorbs* 3,000 J of heat from the surroundings while doing 2,000 J of work on those surroundings? In this case, you have 3,000 J of energy going in and 2,000 J going out. The signs are now easy to understand:

$$\Delta U = -2{,}000 \text{ [work going out]} + 3{,}000 \text{ [heat coming in]} = 1{,}000 \text{ J}$$

In this case, the net change to the system's internal energy is +1,000 J. You can also see negative work when work is done on the system by the surroundings. Say, for example, that a system absorbs 3,000 J at the same time that its surroundings perform 4,000 J of work on the system. You can tell that both of these energies will flow into the system, so its internal energy goes up by 3,000 J + 4,000 J = 7,000 J. If you want to go by the numbers, use this equation:

$$\Delta U = Q - W$$

and then note that because the surroundings are doing work on the system, W is considered negative. So, you get

$$\Delta U = Q - W = 3,000 - -4,000 = 7,000 \text{ J}$$

Examining isobaric, isochoric, isothermal, and adiabatic processes, oh my!

You come across a number of quantities floating around in this chapter — volume, pressure, temperature, and so on — and the way in which these quantities vary as you do work affects the results. For example, if a gas is doing work while you keep its volume constant, the process is different than when you keep its pressure constant.

There are four standard conditions under which work is performed in thermodynamics: Constant pressure, constant volume, constant temperature, and constant heat.

Note that when anything changes in the processes in the following sections, the change is assumed to be *quasi-static,* which means the change comes slowly enough that the pressure and temperature are the same throughout the system's volume.

At constant pressure: Isobaric

When you have a process where the pressure stays constant, it's called *isobaric* ("baric" means pressure). In Figure 15-1, you see a cylinder with a piston being lifted by a quantity of gas as the gas gets hotter. The volume of the gas is changing, but the weighted piston keeps the pressure constant.

What work is the system doing as the gas expands? Work equals F times s, where F represents force and s represents distance, and Force equals P times A, where P is the pressure and A is the area. This means that

$$W = F\Delta s = PA\Delta s$$

But the area times the change in s equals ΔV, the change in volume, so you get

$$W = P\Delta V$$

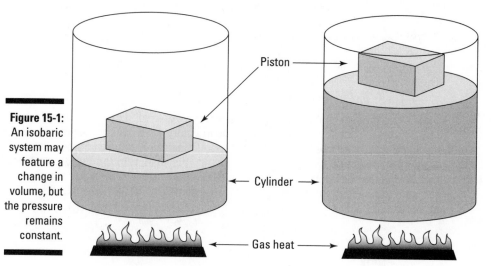

Figure 15-1:
An isobaric
system may
feature a
change in
volume, but
the pressure
remains
constant.

Piston

Cylinder

Gas heat

Graphically, you can see what the isobaric process looks like in Figure 15-2, where the volume is changing while the pressure stays constant. Because $W = P\Delta V$, the work is the area beneath the graph.

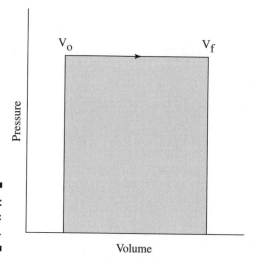

V_0 V_f

Pressure

Figure 15-2:
An isobaric
graph.

Volume

For example, say you have 60 m^3 of an ideal gas at 200 Pa (Pascals; see Chapter 2), and you heat it until it expands (PV = nRT, where *n* represents the number of moles, R represents the gas constant, 8.31, and T represents temperature; see Chapter 14) to a volume of 120 m^3. How much work does the gas do? All you have to do is plug in the numbers:

W = PΔV = (200)(120 − 60) = 12,000 J

The gas does 12,000 Joules of work as it expands under constant pressure.

At constant volume: Isochoric

What if the pressure in a system isn't constant? After all, you don't often come across a setup with a weighted piston, as in Figure 15-1. You're more likely to see a simple closed container, as you see in Figure 15-3, where someone has neglectfully tossed a spray can onto a fire. In this case, the volume is constant, so this is an *isochoric* process. As the gas inside the spray can heats up, its pressure increases, but its volume stays the same (unless, of course, the can explodes).

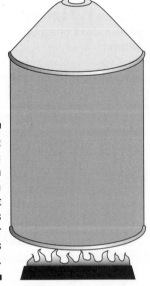

Figure 15-3:
An isochoric system features a constant volume as other quantities vary.

How much work is done on the spray can? Take a look at the graph in Figure 15-4. In this case, the volume is constant, so Fs (force times distance) equals zero. No work is being done — the area under the graph is zero.

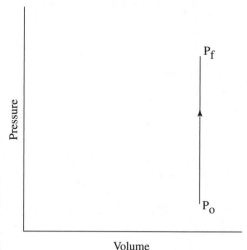

Figure 15-4:
An isochoric
graph.

At constant temperature: Isothermal

In an *isothermal system,* the temperature remains constant as other quantities change. Take a look at the remarkable apparatus in Figure 15-5. It's specially designed to keep the temperature of the enclosed gas constant, even as the piston rises. When you apply heat to this system, the piston rises or lowers slowly in such a way as to keep the product, pressure times volume, constant. Because PV = nRT (see Chapter 14), the temperature stays constant as well.

What does the work look like as the volume changes? Because PV = nRT, the relation between P and V is

$$P = nRT / V$$

You can see this equation illustrated in the graph in Figure 15-6.

The work done shows up in the area underneath the graph. But what the heck is that area? The work done is given by the following equation, where *ln* is the natural log, R is the gas constant, 8.31, V_f represents the final volume, and V_o represents the initial volume:

$$W = nRT \ln(V_f / V_o)$$

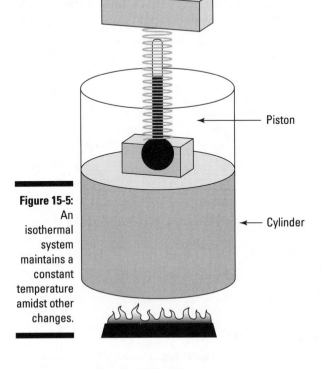

Piston

Cylinder

Figure 15-5:
An
isothermal
system
maintains a
constant
temperature
amidst other
changes.

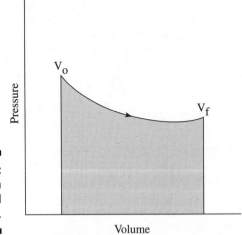

Figure 15-6:
An
isothermal
graph.

Because the temperature stays constant in an isothermal process, and because the internal energy for an ideal gas (see Chapter 14) equals $(\%)nRT$, the internal energy doesn't change. So you have

$$\Delta U = 0 = Q - W$$

In other words,

$$Q = W$$

So, if you immerse the cylinder you see in Figure 15-5 in a heat bath, what would happen? The heat, Q, would flow into the apparatus, and because the temperature of the gas stays constant, all that heat would become work done by the system. For example, say that you have a mole of helium to play around with on a rainy day of temperature 20°C, and for fun you decide to expand it from $V_o = 0.010$ m^3 to $V_f = 0.020$ m^3. What's the work done by the gas in the expansion? All you have to do is plug in the numbers:

$$W = nRT \ln(V_f / V_o) = (1.0)(8.31)(273.15 + 20) \ln(0.020 / 0.010) = 1{,}690 \text{ J}$$

The work done by the gas is 1,690 Joules. The change in the internal energy of the gas is 0 Joules, as always in an isothermal process. And because Q = W, the heat added to the gas is also equal to 1,690 J.

At constant heat: Adiabatic

In an *adiabatic process,* the total heat in a system is held constant. Take a look at Figure 15-7, which shows a cylinder surrounded by an insulating material. The heat in the system isn't going anywhere, so when a change takes place, it's an adiabatic change.

Examining the work done during an adiabatic process, you can say Q = 0, so ΔU (the internal energy change) = –W. Because the internal energy of an ideal gas is U = $(\%)nRT$ (see Chapter 14), the work done is

$$W = (\%)nR(T_o - T_f)$$

where T_o represents the initial temperature and T_f represents the final temperature. So, if the gas does work, it comes from a change in temperature — if the temperature goes down, the gas does work on its surroundings. You can see what a graph of pressure versus volume looks like for an adiabatic process in Figure 15-8. The adiabatic curve in this figure, called an *adiabat,* is different from the isothermal curves, called *isotherms.* The work done when the total heat in the system is constant is the area under the adiabat, as shown in Figure 15-8.

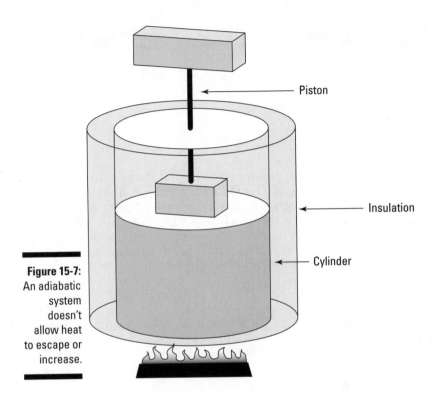

Figure 15-7:
An adiabatic system doesn't allow heat to escape or increase.

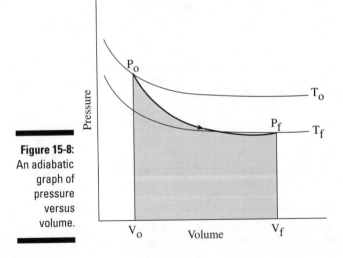

Figure 15-8:
An adiabatic graph of pressure versus volume.

Figuring out specific heat capacities

You can relate the initial pressure and volume to the final pressure and volume this way:

$$P_o V_o^{\gamma} = P_f V_f^{\gamma}$$

What's γ? It's the ratio of the specific heat capacity of an ideal gas at constant pressure divided by the specific heat capacity of an ideal gas at constant volume, c_p / c_v. (The measure of how much heat an object can hold is called its *specific heat capacity;* see Chapter 13 for more.) Allow me to get out the old blackboard and chalk. To figure out specific heat capacities, you need to relate heat, Q, and temperature, T: $Q = cm\Delta T$, where c represents specific heat, m represents the mass, and ΔT represents the change in temperature. In this case, however, it's easier to talk in terms of *molar specific heat capacity*, which is given by C and whose units are Joules/(mole-K). You use a number of moles, *n,* for molar specific heat capacity rather than the mass, *m:*

$$Q = Cn\Delta T$$

How can you solve for C? You must account for two different quantities, C_p (constant pressure) and C_v (constant volume). The first law of thermodynamics (see the previous section in this chapter) states that $Q = \Delta U + W$, so if you can get ΔU and W in terms of T, you're set. The work done is $P\Delta V$, so at constant volume, $W = 0$. And the change in internal energy of an ideal gas is $(\frac{3}{2})nR\Delta T$ (see Chapter 14), so Q at constant volume is

$$Q_v = (\tfrac{3}{2})nR(T_f - T_o)$$

At constant pressure, W isn't 0, because it's $P\Delta V$. And because PV = nRT, you can say $W = P(V_f - V_o) = nR(T_f - T_o)$, so Q at constant pressure is

$$Q_p = (\tfrac{3}{2})nR(T_f - T_o) + nR(T_f - T_o)$$

So, how do you get the molar specific heat capacities from this? You've decided that $Q = Cn\Delta T$, so $C = Q / n\Delta T$. Dividing the above two equations by $n\Delta T$ gives you

$$C_v = (\tfrac{3}{2})R$$
$$C_p = (\tfrac{3}{2})R + R = (\tfrac{5}{2})R$$

Now you have the molar specific heat capacities of an ideal gas. The ratio you want, γ, is the ratio of these two equations:

$$\gamma = C_p / C_v = 5 / 3$$

You can connect pressure and volume at any two points along an adiabat (see the section "At constant heat: Adiabatic") this way:

$$P_o V_o^{5/3} = P_f V_f^{5/3}$$

For example, if you start with 1 liter of gas at 1 atmosphere and end up after an adiabatic change (where no heat is gained or lost) with 2 liters of gas, what would the new pressure, P_f, be? Putting P_f on one side of the equation gives you

$$P_f = P_o V_o^{5/3} / V_f^{5/3}$$

Plugging in the numbers gives you

$$P_f = P_o V_o^{5/3} / V_f^{5/3} = [(1.0)(1.0)^{5/3}] / (2.0)^{5/3} = 0.314 \text{ atmospheres}$$

The new pressure would be 0.314 atmospheres.

When Heat Flows: The Second Law of Thermodynamics

The *second law of thermodynamics* says, formally speaking, that heat flows naturally from an object at a higher temperature to an object at a lower temperature, and it doesn't flow of its own accord in the opposite direction. The law is certainly born out of everyday observation — when was the last time you noticed an object getting colder than its surroundings by itself unless another object was doing some kind of work on it? You can force heat to flow away from an object when it would naturally flow into it if you do some work — as with refrigerators or air conditioners — but it won't happen by itself.

Putting heat to work: Heat engines

You have many ways to turn heat into work. You may have a steam engine, for example, that has a boiler and a set of pistons, or you may have an atomic reactor that generates superheated steam that can turn a turbine. Engines that rely on a heat source to do work are called *heat engines;* you can see how one works by referring to Figure 15-9. A heat source provides heat to the engine, which does work. And the waste heat left over is sent to a *heat sink.* The heat sink could be the surrounding air, or it could be a water-filled radiator, for example. As long as the heat sink is at a lower temperature than the heat source, the heat engine can do work — at least theoretically.

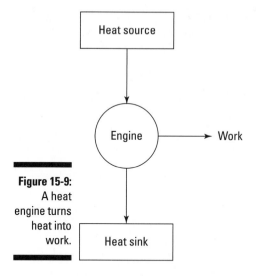

Figure 15-9:
A heat
engine turns
heat into
work.

Evaluating heat's work: Heat engine efficiency

Heat supplied by a heat source is given the symbol Q_h (for the hot source), and heat sent to a heat sink (see the previous section) is given the symbol Q_c (for the cold heat sink). With some calculation, you can find the efficiency of a heat engine. It's the ratio of the work the engine does, W, to the input amount of heat — the fraction of the input heat that it converts to work:

Efficiency = Work / Heat Input = W / Q_h

If all the input heat is converted to work, the efficiency will be 1.0. If no input heat is converted to work, the efficiency is 0.0. Often, the efficiency is given as a percentage, so you express these values as 100 percent and 0 percent. Because total energy is conserved, the heat into the engine must equal the work done plus the heat sent to the heat sink, which means that

$Q_h = W + Q_c$

This means that you can rewrite the efficiency in terms of just Q_h and Q_c:

Efficiency = $W / Q_h = (Q_h - Q_c) / Q_h = 1 - (Q_c / Q_h)$

Say, for example, that you have a heat engine that's 78 percent efficient and can produce 2.55×10^7 Joules to clear off a roadway of water. How much heat does it use, and how much does it reject? Well, you know that $W = 2.55 \times 10^7$ J, and that

Efficiency $= W / Q_h = 2.55 \times 10^7$ J $/ Q_h = 0.78$

That means that

$Q_h = 2.55 \times 10^7$ J $/ 0.78 = 3.27 \times 10^7$ J

The amount of input heat is 3.27×10^7 Joules. So, how much heat gets left over and sent into the heat sink, Q_c? You know that

$Q_h = W + Q_c$

so

$Q_h - W = Q_c$

Plugging in the numbers gives you

$Q_h - W = 3.27 \times 10^7 - 2.55 \times 10^7 = 0.72 \times 10^7 = Q_c$

The amount of heat sent to the heat sink is 0.72×10^7 Joules.

Carnot says you can't have it all

Given the amount of work a heat engine does and its efficiency, you can calculate how much heat goes in and how much comes out (along with a little help from the law of conservation of energy, which ties work, heat in, and heat out together; see Chapter 8). But what about creating 100-percent-efficient heat engines? It would be nice for efficiency if you could convert all the heat that goes into a heat engine into work, but it doesn't work out that way. Carnot says so. Heat engines have some inevitable losses in the real world, such as through friction on the pistons in a steam engine. In the 19th century, an engineer named Sadi Carnot studied this problem, and he came to the conclusion that the best you can do, effectively, is to use an engine that has no such losses.

If the engine experiences no losses, the system will return to the same state it was in before the process took place. This is called a *reversible process*. For example, if a heat engine loses energy overcoming friction, it doesn't have a reversible process, because it doesn't end up in the same state when the process is complete. So, when do you have the most efficient heat engine? When the engine operates reversibly (that is, when the system has no losses). Today, physicists call this Carnot's principle. *Carnot's principle* says that no engine that isn't reversible can be as efficient as a reversible engine, and all reversible engines that work between the same two temperatures have the same efficiency.

Building Carnot's engine

Carnot came up with his own idea of an engine — the *Carnot engine*. His engine operates reversibly, which no real engine can do, so he created a kind of ideal engine. In Carnot's engine, the heat that comes from the heat source is supplied at a constant temperature, T_h. And the rejected heat goes into the heat sink, which is at a constant temperature, T_c. Because the heat source and the heat sink are always at the same temperature, you can say that the ratio of the heat provided and rejected is the same as the ratio of those temperatures (expressed in Kelvins):

$$Q_c/Q_h = T_c/T_h$$

And because the efficiency of a heat engine is:

$$\text{Efficiency} = 1 - (Q_c/Q_h)$$

the efficiency of a Carnot engine is

$$\text{Efficiency} = 1 - (Q_c/Q_h) = 1 - (T_c/T_h)$$

This equation represents the *maximum possible efficiency* of a heat engine. You can't do any better than that. And as the third law of thermodynamics states (see the final section in this chapter), you can't reach absolute zero, which means that T_c is never 0, so you can never have a 100-percent-efficient heat engine.

Using the equation for a Carnot engine

Applying the equation for maximum possible efficiency (see the previous section) is easy. For example, say that you come up with a terrific new invention: a Carnot engine that uses a plane to connect the ground (27°C) as a heat source to the air at 33,000 feet as the heat sink (about –25°C). What's the maximum efficiency you can get for your heat engine? After converting temperatures to Kelvin, plugging in the numbers gives you

$$\text{Efficiency} = 1 - (Q_c/Q_h) = 1 - (T_c/T_h) = 1 - (248.15 / 300.15) = 0.173$$

Your Carnot engine can be only 17.3 percent efficient — not too impressive. On the other hand, assume you can use the surface of the sun (about 5,800 kelvin) as the heat source and interstellar space (about 3.4 kelvin) as the heat sink (such is the stuff science-fiction stories are made of). You'd have quite a different story:

$$\text{Efficiency} = 1 - (Q_c/Q_h) = 1 - (T_c/T_h) = 1 - (3.4 / 5,800) = 0.999$$

You get a theoretical efficiency for your Carnot engine — 99.9 percent.

Going Cold: The Third (and Absolute Last) Law of Thermodynamics

The *third law of thermodynamics* is pretty straightforward — it just says that you can't reach absolute zero through any process that uses a finite number of steps. Which is to say, you can't get down to absolute zero at all. Each step in the process of lowering an object's temperature to absolute zero can get the temperature a little closer, but you can't get all the way there, unless you use an infinite number of steps, which isn't possible.

The strange rewards of near-zero

Although you can't get down to absolute zero with any known process, you can get closer and closer, and if you have some expensive equipment, you discover more and more strange facts about the near-zero world. I have a pal who discovered how liquid helium works at very, very low temperatures. For example, the helium becomes so offbeat that it will climb entirely out of containers by itself if you get it started. For these and some other observations, he and some friends got the Nobel Prize, the lucky dogs.

Part V
Getting a Charge out of Electricity and Magnetism

The 5th Wave   By Rich Tennant

"...finally, those researchers working after hours should limit their investigation to the behavior of protons and electrons, and hereafter refrain from putting eggs in the particle accelerator."

In this part . . .

Two of the most powerful forces in your environment — electricity and magnetism — are invisible. People take them for granted, and if you ask people basic questions about these forces, you see them nod their heads and contemplate how to get away from you. That's what Part V is for — clearing up all the mysteries surrounding electricity and magnetism. You get the full story here, including how the product of electricity and magnetism together — light — behaves with mirrors and lenses.

Chapter 16

Zapping Away with Static Electricity

*E*lectricity is built into everything around you. Each atom has its own charges moving around at incredible speeds. Sometimes, those charges become apparent when you least expect it, like when you get zapped on a dry day by touching a metal doorknob or by closing your car door. Other times, you expect those charges to be around, like when you turn on a light and you know electricity is flowing.

The atmosphere starts getting electric in this chapter, which is about the zap kind of charges — reactions that build up because you have an excess of charge (for example, when you have extra electrons sitting around). That's static electricity. This chapter (and the next) is all about when those charges start moving, which is what you normally think of as electricity. Where you've got electrons, you've got zap, and this chapter features plenty of electrons. I discuss electric charges, electric potential, electric fields, the forces between charges, and plenty more. It all starts with electrons.

Plus and Minus: Electron and Proton Charges

Atoms primarily consist of a nucleus containing charged protons, neutral neutrons, and lightweight electrons that orbit the nucleus at high speeds.

The charged particles — electrons and protons — have the same magnitude of charge, which is

$$e = 1.6 \times 10^{-19} \text{ C}$$

where C is the *Coulomb,* the MKS unit of charge (see Chapter 2). The charge of a proton is $+1.6 \times 10^{-19}$ C and the charge of an electron is -1.6×10^{-19} C (the electron's negative charge is just convention). The electrons, however, are the particles that give you static electricity and the standard electricity that flows through wires. So, if you have 1 entire Coulomb of charge, how many electrons do you have? Because each electron has a charge of magnitude of 1.6×10^{-19} C, you have

$$\text{Total number of electrons} = 1 / 1.6 \times 10^{-19} = 6.25 \times 10^{18} \text{ electrons}$$

You need 6.25×10^{18} electrons to make up 1 Coulomb. But interesting things happen when you get huge numbers of electrons together. Like relatives scattering at the end of a tiresome family reunion, they fly apart.

Push and Pull: Electric Forces

Electric charges exert a force on each other, and if you had 6.25×10^{18} electrons, for example, you'd have a tough time keeping them together. All matter has electric charges, but if an object has an excess number of electrons, it will have a *net negative* charge. And if an object has a deficit of electrons, it will have a *net positive* charge.

As you may know from playing around with magnets, like charges repel, and unlike charges attract. Take a look at Figure 16-1, where some balls on strings have somehow acquired electric charges. The like-charged pairs of balls (++ or −−) will repel each other, but the unlike pairs of charges (+− or −+) will attract each other.

Figure 16-1:
Attractive or
repellant
forces
between
charges.

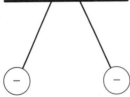 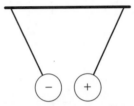 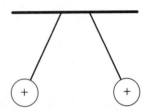

Charging it to Coulomb's law

Charge isn't all positive or negative talk; you have to get numbers involved. How strong are the forces between charged objects? It all depends on how much charge is involved and on how far apart the charges are. In Chapter 5, you see another force that acts between objects — the gravitational force:

$$F = -Gm_1m_2 / r^2$$

where F is the force, G is the universal gravitational constant, m_1 is the mass of one object, m_2 is the mass of the second, and *r* is the distance between them. After you do all the measuring in the lab, the electric force looks like

$$F = kq_1q_2 / r^2$$

In this case, q_1 and q_2 are the two charges acting on each other, measured in Coulombs, and *r* is the distance between them. So, what does *k* represent? A constant whose value is 8.99×10^9 N-m^2/C^2 (you'll always be given this value, so don't worry about memorizing it).

$F = kq_1q_2 / r^2$ is called *Coulomb's law.* It gives you the magnitude of the force between charges. Note that when charges have the same sign, the force between them is positive, which means the charges repel each other; when charges have opposite signs, the force between them is negative, which means the force attracts the two charges.

You sometimes see Coulomb's law written in terms of a constant ε_o, which is called *the permittivity of free space* (the constant has to do with how easily the electric field generated by a charge can extend through space):

$$F = (q_1q_2) / 4\pi\varepsilon_o r^2$$

The constant ε_o has a value of 8.854×10^{-12} C^2/(n-m^2).

Bringing objects together

The distance between objects is an important part of charge and Coulomb's law (see the previous two sections). Say, for example, that you separate two balls 1 meter apart and manage to give each ball 1 Coulomb of charge — one negative, one positive. Now you need to hold them apart — but what's the force pulling them together? Add the numbers to Coulomb's law:

$$F = \frac{kq_1q_2}{r^2} = \frac{(8.99 \times 10^9)(1.0)(-1.0)}{(1.0)^2} = -8.99 \times 10^9 \, N$$

You need 8.99×10^9 Newtons of force to hold the balls apart. That's an unreasonably large number of Newtons — it converts to about 1.1 billion pounds or 560,000 tons, roughly the weight of 10 full oil tankers. You better rethink putting a whole Coulomb of charge on each ball. As you can see, electric forces can get pretty big when you toss whole Coulombs of charge around.

Calculating the speed of electrons

You can relate electrostatic force and centripetal force (see Chapter 10) by looking at the circular orbit of an electron. Imagine that you're sitting around during a physics lecture messing around with a bunch of hydrogen atoms. You know that each atom is made up of a single electron circling a single proton. It's too small to see, but you know the electron is zipping around the proton pretty fast. This realization gets you to thinking — how fast is the electron moving? You know that the electrostatic force is between the electron and the proton, and assuming that the electron's orbit is a circle, the electrostatic force equals the centripetal force (see Chapter 10), so you have Coulomb's law equaling the equation for centripetal force:

$$F = (kq_1q_2) / r^2 = mv^2 / r$$

The mass of an electron is 9.11×10^{-31} kg, and the radius of its orbit is 5.29×10^{-11} m. So, plugging in numbers for the electrostatic force — the constant k and the charges of the electron and proton — gives you

$$F = \frac{kq_1q_2}{r^2} = \frac{(8.99 \times 10^9)(1.6 \times 10^{-19})(-1.6 \times 10^{-19})}{(5.29 \times 10^{-11})^2} = 8.22 \times 10^{-8}\,N$$

You've found the force between the electron and the proton, and that force has to equal the centripetal force, so

$$F = mv^2 / r = (9.11 \times 10^{-31})v^2 / (5.29 \times 10^{-11}) = 8.22 \times 10^{-8}\,N$$

Solving for v gives you 2.18×10^6 meters per second, or about *4.86 million miles per hour.* Try to picture that speed sometime; it's approaching 1 percent of the speed of light.

Looking at forces between multiple charges

Not every charge problem you come across features two charges acting on each other. Whenever you have more than two charges affecting each other, you need to use vectors to find the net force on any one charge. (For more information on vectors, see Chapter 4.)

Take a look at Figure 16-2, which features three charges, all acting on each other — two negative charges and one positive. What's the net force on the positive charge?

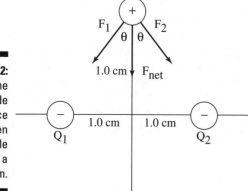

Figure 16-2:
The
magnitude
of the force
between
multiple
charges is a
vector sum.

The forces on the positive charge, Q, due to the two negative charges, Q_1 and Q_2, are shown in Figure 16-2 as F_1 and F_2. The two forces sum up to give you F_{net}. Assuming $Q_1 = Q_2 = -1.0 \times 10^{-8}$ C, $Q = 3.0 \times 10^{-8}$ C, and the charges are at 1.0 cm on the X- and Y-axes, as shown in the figure, what's F_{net}? Using the Pythagorean theorem (see Chapter 2), you know that $\theta = 45°$. In magnitude, $F_1 = F_2$, so

$$F_{net} = 2F_1 \cos(45°) = \sqrt{2}\, F_1$$

What's the magnitude of F_1?

$$F_1 = \frac{kQ_1 Q}{r^2} = \frac{(8.99 \times 10^9)(1.0 \times 10^{-8})(3.0 \times 10^{-8})}{(0.10)^2} = 8.99 \times 10^{-5}\,\text{N}$$

F_1 equals 8.99×10^{-5}N, which means you can find the net force on the positive charge:

$$F_{net} = 2F_1 \cos(45°) = \sqrt{2}\, F_1 = 1.27 \times 10^{-4}\,\text{N}$$

The magnitude of the net force on the positive charge, found as a vector sum (see Chapter 4), is 1.27×10^{-4} N.

Influence at a Distance: Electric Fields

To find the force between charges, you have to know the magnitude (or size) of both charges. But if you have an array of charges, you have an array of magnitudes. What if someone else wants to add a charge to the array? You don't know how much charge the person will add — will the person bring a charge of 1.0 Coulomb to your array? How about 1.0×10^8 C or 1.0×103 C? Who knows?

That's why physicists created the concept of the *electric field* — to describe how your array of charges will work with some other charge. All other people have to do to work with your array of charges is to multiply the electric field at any point (after you tell them what the electric field from your array is) by the charges they have. You express an electric field in Newtons per Coulomb and use the symbol E. (Note that an electric field is a vector quantity, meaning it has a direction and a magnitude; see Chapter 4.) Here's how you define an electric field:

$$E = F / q$$

where F represents force and *q* represents electric charge. In other words, electric field is the force per Coulomb at any particular point. The direction of the electric field at any one point is the force felt by a *positive charge.*

Say, for example, that you're out walking your pet 1.0-Coulomb charge. It's a sunny day, and you're both enjoying the weather, but suddenly and unaccountably, you find yourself in a 5.0-Newton-per-Coulomb electric field pointing in the direction from which you came. You can see the picture in Figure 16-3.

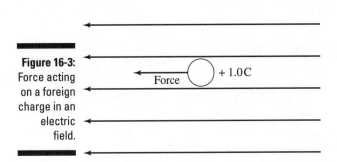

Figure 16-3:
Force acting on a foreign charge in an electric field.

What would happen? Your pet 1.0-C charge would suddenly experience a force opposite to your direction of travel:

$$F = qE = (1.0)(5.0) = 5.0 \text{ Newtons}$$

REMEMBER

If you and your pet 1.0-C charge turn around, the force will be in your new direction of travel. That's how electric fields work — someone tells you what the electric field is, and you can explain how much force would be exerted on a particular charge in that field. The force is in the same direction as the electric field at a particular point if the charge is positive, but if the charge is negative, the force is in the opposite direction.

Because the electric field at any point is a vector (meaning it has a direction and a magnitude), you can find it through vector addition (see Chapter 4 for the specifics of vector addition). You can see this illustrated in Figure 16-4, where two electric fields, one horizontal and one vertical, exist in the same area. The net electric field is their sum.

Figure 16-4:
Adding
electric
fields
produces a
net electric
field.

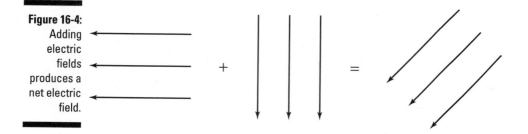

Coming from all directions: Electric fields from point charges

Not all electric fields you encounter look nice and even (like the one you see in Figure 16-3). For example, what's the electric field from a point charge look like? A *point charge* is just a tiny charge. You know that if you have a charge Q, it will create an electric field, but what's that electric field going to be? Well, the equation for an electric field, E = F / q, makes this question fairly easy to figure out. Say, for example, that you have a *test charge* (a charge you use to measure forces with), *q,* and you measure the force from the charge Q at a variety of points. The following is the force you get (using the equation I discuss in the section "Push and Pull: Electric Forces" earlier in this chapter):

$$F = (kqQ) / r^2$$

So, what's the electric field? You divide by the magnitude of your test charge, *q:*

$$E = F / q = kQ / r^2$$

The magnitude of the electric field from a point charge is $E = kQ / r^2$. The electric field is a vector (see Chapter 4), so which direction does it point? To determine the direction of the electric field, go back to your test charge, q, and say that q is a positive charge (remember, an electric field is defined as the force per Coulomb on a positive charge).

Anywhere in the electric field, the force on q from Q is *radial*, or along a line connecting the two charges' centers. When Q is positive, the force on q, which is also positive, will be directed away from Q; therefore, the electric field at any point will also be directed away from Q. You can see this point illustrated in Figure 16-5, which shows an electric field as a collection of electric-field lines, an idea first pioneered by Michael Faraday in the 19th century.

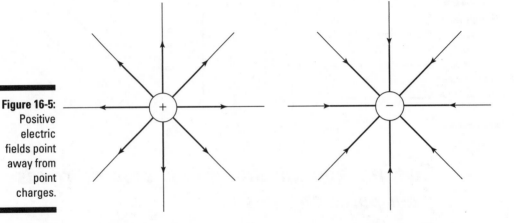

Figure 16-5:
Positive
electric
fields point
away from
point
charges.

Looking at electric-field lines can give you a good idea of what an electric field looks like qualitatively (not quantitatively, in terms of numbers). When the electric-field lines are close together at a point A compared to another point B, the electric field is stronger at A than at B. Note also that the electric field points outward from positive charges and inward toward negative charges, as shown in Figure 16-5.

What if you're asked to examine the electric field from more than one charge? You have to add the electric field as vectors at every point. For example, if you have two point charges — a positive charge and a negative charge — you end up with the electric field you see in Figure 16-6.

Field lines (like the ones you see in Figure 16-6) start from a positive charge and keep going until they end on a negative charge; they don't just start or stop in empty space.

Charging nice and steady: Electric fields in parallel plate capacitors

When dealing with vectors (see Chapter 4), an electric field between multiple point charges isn't the easiest concept to come to grips with (see the previous section for a discussion of multiple point charges). To make your life easier, physicists came up with the *parallel plate capacitor.* A *capacitor* is an object that stores charge in the way that a parallel plate capacitor does — by holding charges separate so that they attract each other, but with no way for them to go from one plate to the other by themselves.

Take a look at Figure 16-7. Here, a charge +*q* is spread evenly over one plate, and a charge –*q* is spread evenly over another. That's great for your purposes, because the electric field from all the point charges on these plates cancels out all components except the ones pointing between the plates. In other words, parallel plate capacitors give you constant electric fields, all in the same direction, which are easier to work with than fields from point charges.

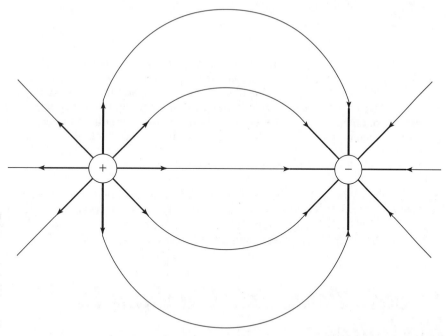

Figure 16-6:
Adding
electric
fields from
two point
charges.

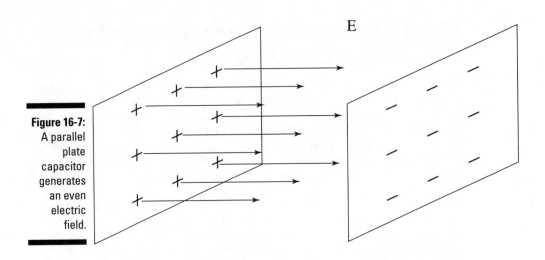

If you do a lot of math, you can figure out that the electric field, E, between the plates is constant (as long as the plates are close enough together), and in magnitude, it's equal to

$$E = q / \varepsilon_o A$$

where ε_o is the permittivity of free space, 8.854×10^{-12} C^2/(n-m^2) (see the section "Charging it to Coulomb's law" earlier in this chapter), q is the total charge on either of the plates (one plate has charge $+q$ and the other has $-q$), and A is the area of each plate. You can also write the equation in terms of the *charge density*, σ, on each plate, where $\sigma = q / A$ (the charge per square meter). Here's what including the charge density makes the equation look like:

$$E = q / \varepsilon_o A = \sigma / \varepsilon_o$$

Working with a parallel plate capacitor makes life a little easier because the electric field has a constant value and a constant direction (from the positive plate to the negative plate), so you don't have to worry about where you are between those plates to find the electric field.

Electric Potential: Cranking Up the Voltage

Electric fields (see the previous section) are only part of the electric story; you have other concepts to take into account. You're dealing with forces in electricity, so you have to address the idea of *potential energy,* or the energy stored in an object or system. Mixing force and potential energy is a natural fit; for example, when you lift a weight in a gravitational field, you end up with potential energy, the energy stored in the object because of its new position:

$$PE = mgh_f - mgh_o$$

where m represents mass, g represents the acceleration due to gravity, h_f represents the final height, and h_o represents the initial height. Because a force acts on the charges in an electric field, you can also speak about potential energy in electric fields. Such potential energy is *electric potential energy,* and a change in electric potential energy creates a new quantity called *voltage,* or the driving force of electric current.

Calculating electric potential energy

Electric potential energy is the potential energy stored in electric fields. Introducing the concept of energy also introduces the concept of work (see Chapter 8). Assume that you're moving the positive charge in Figure 16-8 toward the positively charged plate to the left. Which way would the force be on that charge? You can think of the electric field as coming from positive charges and going to negative charges, so the charge in the figure is facing a whole sheet of positive charges. And because the charge itself is positive, the force on it will push it away from the other positive charges, so the charge will be directed toward the right. The sheet of negative charge behind the positive charge will also be pulling it to the right.

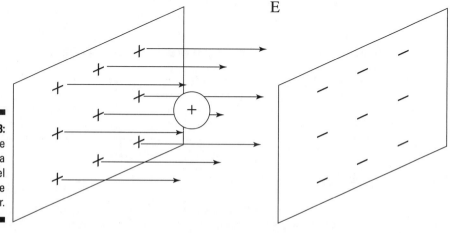

Figure 16-8:
A positive charge in a parallel plate capacitor.

So, how much potential energy does a positive charge acquire as it moves between the plates — going from right to left and opposing the force that would pull it back? The work you do on the charge must be equal to its gain in potential energy, and that work equals

$$W = Fs$$

where F represents force and *s* represents distance. The force on the positive charge is qE, where *q* is the magnitude of the charge and E is the electric field the charge is in. So, the equation becomes

$$W = qEs$$

This equation gives you the magnitude of the potential energy the charge acquires. When the electric field is constant in direction and strength, you can say that

$$Potential \ Energy = qEs$$

 If you set the potential energy of the charge as 0 at the negative plate (as you may set the gravitational potential energy of a ball to 0 when it's resting on the ground), the change, Δ, in its potential energy as you move toward the positive plate is

$$\Delta PE = qE\Delta s$$

As you define electric field as the force per Coulomb (see the section "Influence at a Distance: Electric Fields" earlier in this chapter), so physicists like to define a new quantity for electric potential energy, which is the potential energy per Coulomb. The new quantity created by the change in electric potential energy is *voltage*.

Realizing the potential in voltage

Physically speaking, *voltage* is called *electric potential* (as opposed to electric potential energy). You also call it just *potential*. It's the quantity created by a change in potential energy; you measure it in volts, with the symbol V (volts are the same as Joules per Coulomb; see Chapter 2). The electric potential, V, at a particular point is the electric potential energy of a test charge divided by the magnitude of the test charge:

$$V = PE / q$$

In other words, voltage is the potential energy per Coulomb. The work to move a charge, *q,* from a negative plate to a distance, *s,* toward a positive plate in a parallel plate capacitor (see the section "Charging nice and steady: Electric fields in parallel plate capacitors") is

$$W = qEs$$

That work becomes the charge's potential energy, so the potential at that location is

$$V = PE / q = Es$$

Say, for example, that you happen to drive by a car pulled over on the side of the road with its hood up. You stop and ask what's wrong. "Darn car won't go," the driver says.

That doesn't exactly pinpoint the problem, you think, as you get out your voltmeter and test the car's battery — 12.0 volts, so no problem there. Given that 12.0 volts is the change in potential energy per Coulomb going from one terminal of the battery to the other, how much work does it take to move one electron between those terminals? You know that

$$V = W / q$$

so

$$W = qV$$

You get out your calculator as the stranded driver watches with interest. Remembering that the magnitude of the charge of an electron is 1.6×10^{-19} C (see the section "Plus and Minus: Electron and Proton Charges" earlier in this chapter), you plug in the numbers:

$$W = qV = (1.6 \times 10^{-19})(12.0) = 1.92 \times 10^{-18} \text{ J}$$

"It takes," you say proudly, "1.92×10^{-18} Joules to move one electron between the terminals of your battery."

The driver loses the hopeful expression and looks at you strangely.

Discovering that electric potential is conserved

When it comes to working with kinetic and potential energy (check out a discussion of these topics in Chapter 8), you can always rely on the conservation of total energy:

$$E_1 = PE_1 + KE_1 = PE_2 + KE_2 = E_2$$

You may be happy to find out that energy is conserved in electric potential energy, too. Say, for example, that an unfortunate piece of dust, mass 1.0×10^5 kg, hits the negative plate in a parallel plate capacitor (see the section "Charging nice and steady: Electric fields in parallel plate capacitors") and gets a charge of -1.0×10^5 C. Negatively charged, the dust piece finds that it can't resist the pull from the positive plate and starts drifting toward it.

The voltage difference between the two plates is 30 V — how fast will the speck of dust be traveling when it hits the positive plate (neglecting air resistance)? Because energy is conserved, the potential energy the charge has on the negative plate will be converted into kinetic energy ($KE = \frac{1}{2}mv^2$) when it hits the positive plate. You find the magnitude of the potential energy the dust mote has to start with the equation

$PE = qV$

Plugging in the data gives you

$PE = qV = (1.0 \times 10^5)(30) = 3.0 \times 10^4$ J

This energy turns into kinetic energy, so you get

$KE = \frac{1}{2}mv^2 = 3.0 \times 10^4$ J

Plugging the numbers into this equation gives you

$KE = \frac{1}{2}(1.0 \times 10^5)v^2 = 3.0 \times 10^4$ J

Solving for v, the dust mote ends up with the following speed:

$v = 7.75$ m/s

Which means the dust mote will impact the positive plate at a speed of about 7.75 meters per second, or about 17 miles per hour.

Finding the electric potential of point charges

The electric potential, or voltage (V; see the preceding section), between the plates of a capacitor depends on how far you are from the negative plate toward the positive plate, s (see the section "Charging nice and steady: Electric fields in parallel plate capacitors" for more on capacitors):

$V = Es$

Now think about a point charge (a tiny charge) Q — the electric field isn't as constant as it is between the plates of a capacitor. How can you figure out the potential at any distance from the point charge? The force on a test charge, q, is equal to

$F = kQq / r^2$

where k represents a constant whose value is 8.99×10^9 N-m^2/C^2 and r represents distance. You also know from the section "Coming from all directions: Electric fields from point charges" that the electric field at any point around a point charge, Q, is equal to

$$E = kQ / r^2$$

So, how do you find the electric potential for point charges? At an infinite distance away, the potential is zero. As you bring a test charge closer, to a point r away from the point charge, you simply have to add up all the work you do and then divide by the size of the test charge. The result turns out to be gratifyingly easy; here's what the equation looks like:

$$V = \frac{W}{q} = kQ / r$$

This is the potential, measured in volts, at any point a distance r from a point charge of charge Q, where zero potential is at $r = \infty$. This makes sense if you keep in mind that potential is the work it takes to get a test charge to a certain location divided by the size of the test charge. For example, say you have a proton, $Q = +1.6 \times 10^{-19}$C, at the center of a hydrogen atom, and that the typical orbit of an electron is 5.29×10^{-11} meters away from the proton. What's the voltage from the proton at that distance? You know that

$$V = kQ / r$$

Plugging in the numbers gives you

$$V = \frac{kQ}{r} = \frac{(8.99 \times 10^9)(1.6 \times 10^{-19})}{5.29 \times 10^{-11}} = 27.2 \text{ volts}$$

The electric potential at that distance to the proton is a full 27.2 volts. Not bad for a tiny point charge.

You can also represent electric potential graphically, just as you can represent electric fields with field lines. You can display potential by using *equipotential surfaces* — surfaces where the potential on each surface is the same. For example, because the potential from a point charge depends on distance (or the radius of a sphere), the equipotential surfaces due to a point charge are spheres around that point charge, as you see in Figure 16-9.

What about equipotential surfaces between the plates of a parallel plate capacitor? If you start at the negatively charged plate and move a distance s toward the positively charged plate, you know that

$$V = Es$$

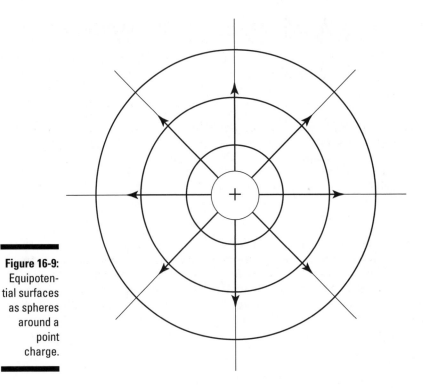

Figure 16-9:
Equipoten-
tial surfaces
as spheres
around a
point
charge.

In other words, the equipotential surfaces depend only on how far you are between the two plates, as you can see in Figure 16-10, where two equipotential surfaces are between the plates.

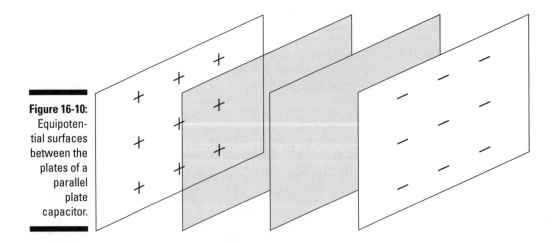

Figure 16-10:
Equipoten-
tial surfaces
between the
plates of a
parallel
plate
capacitor.

Getting fully charged with capacitance

A *capacitor* stores charge by holding charges separate so that they attract each other but don't have a way to go from one plate to the other by themselves.

How much charge is stored? That depends on the *capacitance*, C, of the capacitor. The amount of charge that appears on both plates of the capacitor is equal (the charges are opposite in sign) and depends on the voltage between those plates, as given by the following equation, where C is the capacitance:

$$q = CV$$

where q is an electric charge. For a parallel plate capacitor, the electric field, E, equals the following (see the section "Influence at a Distance: Electric Fields"):

$$E = q / \varepsilon_o A$$

where ε is a constant and A represents the area of the plates. And the voltage between the plates separated by a distance s is

$$V = Es$$

Therefore,

$$V = qs / \varepsilon_o A$$

Because q = CV, you can solve the preceding equation for q / v to get

$$C = q / V = \varepsilon_o A / s$$

The equation $C = q / V = \varepsilon_o A / s$ allows you to find the capacitance for a parallel plate capacitor whose plates each have area A and are a distance s apart. The MKS unit for capacitance is Coulombs per volt, also called the *Farad*, F.

Good stuff, but you're not done yet. Most capacitors don't depend on just air between the plates — they use a dielectric between the plates. A *dielectric* is a semi-insulating material that increases how much charge the capacitor can hold by its dielectric constant, κ. So, when the space between the plates of a parallel plate capacitor is filled with a dielectric of dielectric constant κ, the capacitance increases to

$$C = \kappa \varepsilon_o A / s$$

For example, the dielectric constant of mica (a mineral commonly used in capacitors) is about 5.4, so it increases the capacitance of a capacitor to 5.4 times that of the same capacitor with a vacuum between the plates, because a vacuum has a dielectric constant of 1.0.

Because a capacitor is filled with charges that you've separated from each other, it has some energy associated with it. When you charge a capacitor, you assemble the final charge, q, in an average electric potential, V_{avg} (you use the average potential because the potential increases as you add more charge, so over time, each bit of charge you bring to the plates interacts with, on average, the average potential), so the energy stored is

$$\text{Energy} = qV_{avg}$$

What's V_{avg}? Because the voltage is proportional to the amount of charge on the capacitor ($q = CV$), V_{avg} is half the final charge:

$$V_{avg} = \tfrac{1}{2}V$$

Plugging this equation for V_{avg} into the previous equation for energy and substituting $q = CV$ finally gives you

$$\text{Energy} = \tfrac{1}{2}CV^2$$

You can now find the energy stored in a capacitor: $E = (\tfrac{1}{2})\,CV^2$. When you plug values into this equation, the results come out in Joules, J.

Chapter 17

Giving Electrons a Push with Circuits

....

....

Static electricity occurs when you have an excess or deficit of electrons hanging around, which means you have negatively or positively charged objects. In the normal kind of electricity — the kind that flows through electrical circuits — you don't have an excess of charge, which means no net charge. Instead, voltage, like the voltage you may get from a battery or wall outlet, provides an electric field in conductive wires, and electrons flow in response to that electric field. (For more on voltage, see Chapter 16.)

This chapter is all about electrons in motion in circuits — the kind of circuits you're familiar with. Chapter 16 discusses static electricity, and this chapter is all about standard electricity. Here, I explore the differences through discussions of currents, Ohm's law, electrical power, and finally, circuits.

Electrons on the March: Current

When electrons are flowing, a current exists. But how do you get electrons to start moving to create electrical current? You have to provide an electromotive force, or an EMF. An EMF gives you a difference in potential (or voltage), which means that the electrons will feel a force.

So, what exactly is an EMF? A battery. Or a wall socket. An *EMF* is anything that can provide a voltage, because a voltage is all it takes to create an electric field in a wire, and that electric field gets the electrons moving (you find out in Chapter 16 that Electric field = F [force] / q, [electric charge]). Like the idea of voltage from Chapter 16, there may be a lot about current that's familiar to you.

In physics, electric current is given the symbol I, and its units are *amperes,* or just plain *amps*. The symbol for amps is A.

Defining current

How do you actually define *current?* It's the amount of charge passing through part of a circuit in a certain amount of time. Here's how that explanation looks in equation form:

$$I = q / t$$

where q is electric charge and t is time. If you have 1 Coulomb of charge passing per second, you have 1 amp of current.

Calculating the current in batteries

You can calculate the current flowing through a battery if you know the amount of charge in the circuit and the time, $I = q / t$. Take a look at Figure 17-1; the two vertical lines at the top represent a battery (the lines are supposed to be reminiscent of metal plates of various kinds, which were alternated with chemicals and stacked to create the early forms of batteries).

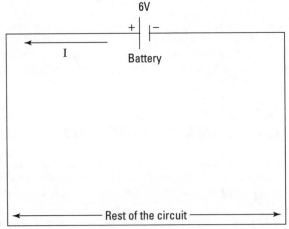

Figure 17-1:
A battery at
work.

The battery provides 6.0 volts of electromotive force (EMF), which drives the current around the circuit. If you have 19.0 Coulombs of charge flowing around the circuit in 30 seconds, what's the current?

$$I = q / t = 19.0 / 30 = 0.633 \text{ amps}$$

In this case, you have 0.633 amps flowing in the circuit. Note that the current, I, comes out of the positive side of the battery, which is the longer line in a battery symbol. If you have only one battery providing the electromotive force in a circuit, the current will flow out of its positive side (the long side) and into its short side (the negative side).

It helps if you think of a battery in a circuit as a voltage step; in other words, the current that arrives at the negative side of the battery is given a lift up by the battery — 6.0 volts in the case of Figure 17-1 — and then starts flowing through the circuit again.

Although current is always shown traveling through a circuit from the positive to the negative signs of a battery, the truth shows that the electrons actually flow in the opposite direction. Why the difference? Historically, early researchers thought that positive charges flowed around a circuit, but the opposite turns out to be true. That's not a problem, however, as long as you're consistent about what you do and always have current flowing out of the positive side of a battery (unless you have another battery or voltage source around that overpowers the first battery).

Giving You Some Resistance: Ohm's Law

Resistance is what ties together the amount of voltage you apply to the current you actually get. One equation ties voltage, current, and resistance together:

$$V = IR$$

where V is voltage, I is current, and R is resistance. Resistance is measured in ohms, and the symbol for ohms is the Greek letter omega: Ω. So, if you apply a certain voltage, V, across a certain resistance, R, you get a current, I. This is called *Ohm's Law,* after its discoverer, Georg Simon Ohm (he made the discovery in the 19th century).

Determining current flow

Using Ohm's law, you can determine how much current will be flowing from the positive to the negative terminal of a battery. Take a look at the circuit you see in Figure 17-2, where a battery of 6.0 V is providing the current that flows through a resistor, R, of 2.0 Ω.

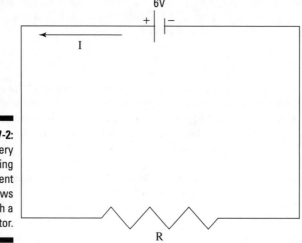

Figure 17-2:
A battery
providing
the current
that flows
through a
resistor.

Using Ohm's law, you know that

$$I = V / R$$

Putting in the numbers gives you

$$I = 6.0 / 2.0 = 3.0 \text{ A}$$

You get a current of 3.0 amps, flowing counterclockwise around the circuit.

Examining resistivity

In your electric travels, you may come across a quantity called *resistivity*, ρ, whose units are Ω-m. If you have a current flowing through a material, the resistivity of that material tells you how much resistance you'll have. Physicists have calculated the resistivities of many common materials for you, and I've laid some of them out in Table 17-1.

You can find the resistance by multiplying the resistivity by the length, L, of the material (the longer it is, the more resistance it causes) and dividing by the material's cross-sectional area, A (the more area the current has for flow, the lower the resistance):

$$\text{Resistance} = \rho L / A$$

Table 17-1	Resistivities of Common Materials
Material	**Resistivity**
Copper	1.72×10^{-8} Ω-m
Rubber	1.0×10^{15} Ω-m
Aluminum	2.82×10^{-8} Ω-m
Gold	2.44×10^{-8} Ω-m
Wood	3.0×10^{10} Ω-m
Carbon	3.5×10^{-5} Ω-m

Powering Up: Wattage

Some household items, like light bulbs or hair dryers, use electrical power. Power is measured in *watts*. How do you calculate electrical power? The amount of work done to move a charge, *q,* around a circuit is qV, where V is the electromotive force. If you divide that work by time, you get power:

$$P = W / t = qV / t$$

However, the charge, *q,* divided by the time, *t,* equals the current, I, so

$$P = W / t = qV / t = IV$$

The power provided to the circuit by an EMF source, like a battery, is P = IV. For example, assume a battery provides 0.5 amps to a light bulb at 10 volts. What's the power of the bulb? P = IV, so the power is (0.5)(10) = 5 watts. However, because I = V / R, you can get variations. The power supplied to a circuit by a particular voltage is

$$P = IV = V^2 / R = I^2R$$

Flowing from One to the Other: Series Circuits

In the previous sections in this chapter, you deal with current flow in a single resistor; however, you can have two resistors in a circuit, too, as shown in Figure 17-3.

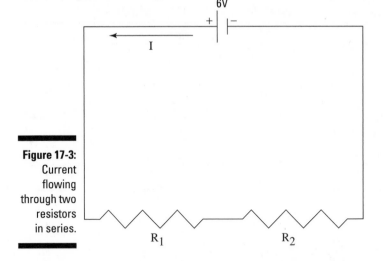

Figure 17-3:
Current
flowing
through two
resistors
in series.

Two resistors can be connected in *series,* where the current in the circuit flows through one and then the other on its way back to the electromotive force (EMF; see the first section in this chapter). How do you handle a case where you have two resistors in series where the current has to flow through *both,* R_1 and R_2, before going back to the battery? The total resistance, R_T, must be the sum of the two resistances:

$$R_T = R_1 + R_2$$

To get the total resistance of two resistors in series, R_1 and R_2, you add them together. For example, if $R_1 = 10 \ \Omega$ and $R_2 = 20 \ \Omega$, and the battery is 6.0 V, what's the current flowing in the circuit? The total resistance must be 30 Ω, so

$$I = V / R = 6.0 / 30 = 0.2 \text{ amps}$$

Splitting the Current: Parallel Circuits

If multiple resistors work in the same circuit, they don't have to be series resistors (see the previous section) where the current goes from one to the other. Two resistors, R_1 and R_2, can be connected so the current splits, as in Figure 17-4. Some of the current goes through one resistor, and the rest goes through the other.

The resistors in Figure 17-4 are *parallel,* which means that they have the same voltage applied across them, not the same current going through them.

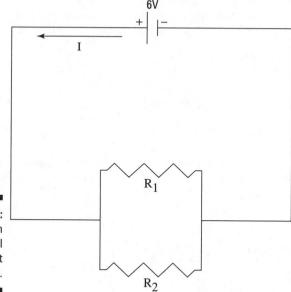

Figure 17-4:
Resistors in
a parallel
circuit split
the current.

The voltage across each parallel resistor is the same — the 6.0 V of the
battery — which is the difference between series and parallel resistors.
When the resistors are in series, the same current flows through them. When
they're parallel, they have the same voltage across them.

So, what's the total resistance when R_1 and R_2 are parallel? The total current,
I_T, is the current that flows through both resistors, $I_1 + I_2$:

$$I_T = I_1 + I_2$$

And because $I = V / R$ (see the section "Giving You Some Resistance:
Ohm's Law"), you can write

$$I_T = I_1 + I_2 = V_1 / R_1 + V_2 / R_2$$

The whole trick to parallel circuits is that $V_1 = V_2$, so if you call the voltage V,
you can say that

$$I_T = I_1 + I_2 = V_1 / R_1 + V_2 / R_2 = V(1 / R_1 + 1 / R_2)$$

This has the form $I_T = V / R_T$, so you finally find that

$$1 / R_T = 1 / R_1 + 1 / R_2$$

This equation tells you how to calculate the total resistance of two resistors in a parallel circuit. If you continue the derivation for any number of resistors, you get the following way of calculating the total resistance:

$$1 / R_T = 1 / R_1 + 1 / R_2 + 1 / R_3 + 1 / R_4 \ldots$$

For example, if $R_1 = 10\ \Omega$ and $R_2 = 30\ \Omega$ in Figure 17-4, and the battery is 6.0 V, what's the current that's flowing? The total resistance in the circuit is

$$R_T = 1 / 10 + 1 / 30 = 4 / 30$$

To find total resistance in parallel circuits, you add the reciprocals of the resistances and take the reciprocal of the result. Therefore, total resistance is $^{30}\!/_4\ \Omega$, making the current $(6.0) / (^{30}\!/_4) = 0.8$ A.

Looping Together Electricity with Kirchoff's Rules

Unfortunately, you can't always break circuits down into series and parallel components, which is why *Kirchoff's rules,* named after their inventor, Gustav Kirchoff, are important. His rules allow you to analyze circuits of varying complexity by applying two simple principles:

- ✔ **The junction rule:** The total current going into any point in a circuit must equal the total current going out of that point.

- ✔ **The loop rule:** Around any closed loop in a circuit, the sum of the potential rises (from a battery, for example) must equal the sum of the potential drops (from a resistor, for example).

The junction rule is pretty easy to understand — the current into any particular point must equal the current out of that point. But what about the loop rule, which says that the rises in potential around any closed loop must equal the drops in potential?

The loop rule means that the electrons going around the circuit are lifted up as far as they drop, and they end where they started. Batteries, for example, can be potential rises — when an electron enters the negative side and comes out of the positive side, the voltage of the battery has raised it. On the other hand, when an electron enters a resistor, it takes some effort to get it through the resistor (which is why a resistor is called a resistor), so there's a drop in potential when the electron comes out the other side.

Implementing the loop rule

You can see an example that illustrates the loop rule in Figure 17-5. You see two resistors and two batteries. What's the current flowing in this loop?

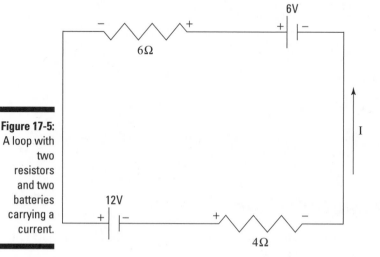

Figure 17-5:
A loop with two resistors and two batteries carrying a current.

Kirchoff's loop rule says

$\Sigma V = 0$ around a loop

where ΣV is the sum of the voltages around a loop. How can you put this rule to work?

To attack this kind of problem, choose a current direction by drawing an arrow for I, as you see in Figure 17-5. The actual current may end up going in the other direction, but that's okay, because then the value of the current you get will be negative, showing you that the actual current flows the other way. After you decide on the direction of the current — counterclockwise in this example — it helps to draw + signs where the current goes into a resistor and – signs were it goes out (these steps aren't part of Kirchoff's rules; I'm just providing technique that I've found helpful).

You know that $\Sigma V = 0$ around the loop and that the potential drop across a resistor is $V = IR$, so you're all set. Just go around the loop in one direction (clockwise or counterclockwise, it doesn't matter), and when you encounter a + or – sign (either on a resistor or a battery), put that sign down, followed by the potential drop or rise. For example, starting at the 6-volt battery and going clockwise, you get the following equation from Kirchoff's loop rule:

$+6 - 4I - 12 - 6I = 0$

When you combine terms, you get

$$6 - 12 = -6$$

$$-4I + -6I = -10I$$

so

$$-6 - 10I = 0$$

or

$$I = -6 \text{ amps} / 10$$

The current is –0.6 amps.

REMEMBER

The fact that the current is negative indicates that it actually goes in the opposite direction of the arrow I choose in Figure 17-4.

Using multiple-loop circuits

Kirchoff's rules seem wasted to me when used on only one-loop circuits. Take a look at the new challenge you see in Figure 17-6.

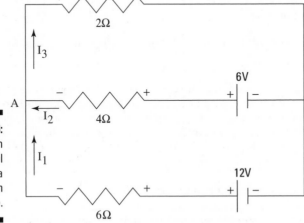

Figure 17-6:
Resistors in
parallel
share a
common
voltage.

The figure shows three branches of this circuit and three different currents — I_1, I_2, and I_3. Can you find these currents? Sure, but you need both of Kirchoff's rules here. The junction rule says that $\Sigma I = 0$ at any point, where ΣI is the sum of currents flowing into and out of that point, so take a look at point A at the left in Figure 17-6. I_1 and I_2 flow into it, and I_3 flows out of it, so

$$I_1 + I_2 = I_3$$

Now for the loop rule, which says $\Sigma V = 0$. The example has three loops: the two internal loops and the external, overall loop. Because there are three unknowns — I_1, I_2, and I_3 — all you need are three equations, and the $\Sigma I = 0$ rule has already given you one. So, you go around the two internal loops to get these equations. From the top loop, you get

$$+6 - 2I_3 - 4I_2 = 0$$

And from the bottom loop, you get

$$+12 - 6 + 4I_2 - 6I_1 = 0$$

There you have it; three equations in three unknowns:

$$I_1 + I_2 = I_3$$
$$+6 - 2I_3 - 4I_2 = 0$$
$$+12 - 6 + 4I_2 - 6I_1 = 0$$

If you substitute the top equation for I_3 into the second equation, you get

$$+6 - 2(I_1 + I_2) - 4I_2 = 0$$
$$+12 - 6 + 4I_2 - 6I_1 = 0$$

or

$$+6 - 2I_1 - 6I_2 = 0$$
$$+12 - 6 + 4I_2 - 6I_1 = 0$$

You can solve for I_1 in terms of I_2 by using the first equation here:

$$I_1 = 3 - 3I_2$$

You can substitute this value of I_1 in the second equation to get

$$+12 - 6 + 4I_2 - 6(3 - 3I_2) = 0$$

or

$$-12 + 22I_2 = 0$$

So,

$$I_2 = {}^{12}\!/_{22} = {}^6\!/_{11} \text{ amps}$$

You now have one of the currents: $I_2 = {}^6\!/_{11}$ amps. Now you can plug this fraction into

$$+6 - 2I_3 - 4I_2 = 0$$

to get

$$+6 - 2I_3 - 4({}^6\!/_{11}) = 0$$

Or, dividing by 2,

$$+3 - I_3 - {}^{12}\!/_{11} = 0$$

For I_3, you get

$$I_3 = {}^{21}\!/_{11} \text{ amps}$$

Now you have I_2 and I_3. How about I_1? You know that

$$I_1 + I_2 = I_3$$

which means that

$$I_1 = I_3 - I_2$$

Easy math from here on:

$$I_1 = {}^{21}\!/_{11} - {}^6\!/_{11} = {}^{15}\!/_{11}$$

You have all the currents now, thanks to Kirchoff's rules. $I_1 = {}^{15}\!/_{11}$ amps, $I_2 = {}^6\!/_{11}$ amps, and $I_3 = {}^{21}\!/_{11}$ amps.

Finding the currents for this type of problem takes plenty of work and a fair amount of math, but if you can master this, you're all set for loop circuits.

Conquering Capacitors in Parallel and Series Circuits

Resistors aren't the only objects you can place in series or parallel circuits. You can do the same with other electrified pals known as *capacitors* (see Chapter 16). A capacitor is an object that stores charge in the way that a parallel plate capacitor does — by holding charges separate so that they attract each other, but with no way for them to go from one plate to the other by themselves. To add capacitors together in parallel circuits, you just add the capacitance of each capacitor to get the total capacitance:

$$C_T = C_1 + C_2 + C_3 + \ldots$$

Capacitors in parallel circuits

In a situation where the capacitors are parallel, the battery supplies the voltage, and that voltage is the same across both capacitors. Take a look at the situation in Figure 17-7, which shows two capacitors in a parallel circuit.

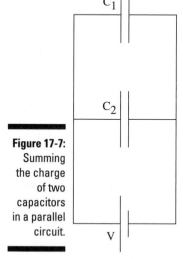

Figure 17-7:
Summing the charge of two capacitors in a parallel circuit.

How do you handle a situation like this? Examine the total charge, Q, stored on those two capacitors, $C_1 + C_2$, which is the sum of the charge stored on each capacitor:

$$Q = C_1V + C_2V$$

Because the battery supplies the same voltage, V, across both capacitors, the voltage across the capacitors is equal, so you can rewrite the equation in terms of a single capacitor whose capacitance is $C_1 + C_2$:

$$Q = (C_1 + C_2)V$$

In other words, if you think in terms of one capacitor, C_T, of capacitance $C_1 + C_2$, you get the same picture:

$$Q = (C_1 + C_2)V = C_T V$$

Capacitors in series circuits

When capacitors are parallel, a battery supplies the voltage, and that voltage is the same across both capacitors. In a series circuit, the *charge* is the same for each capacitor. Take a look at Figure 17-8, which features two capacitors in a series circuit. How do you handle this situation?

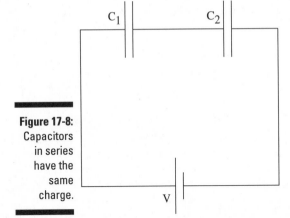

Figure 17-8:
Capacitors
in series
have the
same
charge.

As you see in Figure 17-8, the right-most plate of C_1 and the left-most plate of C_2 are connected to each other but not to the rest of the circuit. In other words, the two plates are isolated from the rest of the circuit, and they start out electrically neutral (total net charge = 0).

Any negative charge, $-q$, that ends up on the right-most plate of C_1 must be equal in magnitude to any positive charge, q, that ends up on the left-most plate of C_2, because the total charge over these two plates has to be zero. And because the total charge over the two plates of a single capacitor must also be zero, the right-most plate of C_1 must have charge q, and the right-most plate of C_2 must have charge q. Therefore, every plate has the same *magnitude* of charge (whether it's positive or negative), q.

Okay, so each capacitor has the same amount of charge. Where does that get you? You now know the total voltage across the two capacitors is equal to

$$V = q / C_1 + q / C_2$$

Because the charge on each capacitor is the same, the equation becomes

$$V = q / C_1 + q / C_2 = q(1 / C_1 + 1 / C_2)$$

Writing the equation in terms of a single total capacitance, C_T, gives you

$$V = q / C_1 + q / C_2 = q(1 / C_1 + 1 / C_2) = q / C_T$$

In other words, to add capacitances in series, you add them like you add parallel resistors (see the section "Splitting the Current: Parallel Circuits") — by adding the reciprocals and taking the reciprocal of the result:

$$1 / C_T = 1 / C_1 + 1 / C_2$$

If you have to deal with more capacitors, you add them together the same way:

$$1 / C_T = 1 / C_1 + 1 / C_2 + 1 / C_3 + \ldots$$

Putting Together Resistors and Capacitors: RC Circuits

I deal with resistors (an electronic component that impedes the flow of current in an electric circuit) and capacitors (objects that store charge by holding charges separate so that they attract each other, but with no way for them to go from one plate to the other by themselves) as separate entities in the previous sections of this chapter — now comes the time to put them together. Take a look at the resistor and capacitor in Figure 17-9. An electronics crew has charged up the capacitor so that it has a voltage, V_o, across it. When the electronics crew closes the switch, they expect a steady current to flow.

However, the current the crew actually gets looks like the graph you see in Figure 17-10. The current starts off at V_o / R as it should (where R represents the resistance), but then it tapers off. What's happening?

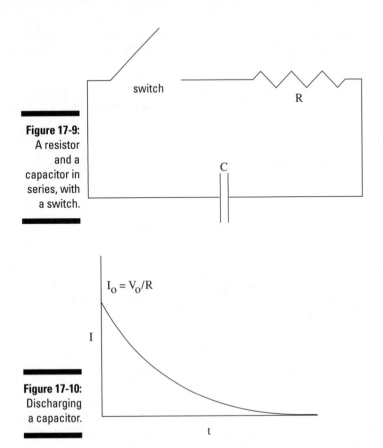

Figure 17-9:
A resistor
and a
capacitor in
series, with
a switch.

Figure 17-10:
Discharging
a capacitor.

The charge is draining off the capacitor in time, so the current diminishes. A capacitor isn't a battery, despite what the electronics crew may want. It can supply current, but only while it still has some charge left. The current starts off at a value of V_o / R, because the capacitor has voltage V_o, and the current it creates flows through a resistor R. But in time, that current decays following this equation:

$$I = I_o e^{-t/RC} = (V_o e^{-t/RC}) / R$$

Here, I is the current, e is the natural log base, 2.71828 (you can probably find a key for e^x on your calculator), t is time, R is the resistance, and C is the capacitance. The charge on the capacitor also follows the same kind of curve:

$$q = q_o e^{-t/RC}$$

Chapter 18

Magnetism: More than Attraction

....

In This Chapter

▶ Running through the magnetic field

▶ Catching up with moving charges

▶ Identifying forces due to magnetic fields

▶ Examining charged particles in a magnetic field

▶ Flowing with the current in magnetic fields

▶ Using solenoids to achieve a uniform magnetic field

....

You can observe a strong connection between electricity and magnetism in the form of moving charges that can create magnetic fields (as in electromagnets and electric motors) and moving magnets that can create electric current (as in electric generators). Even electrons, in their speedy orbits in the atoms of an object, generate magnetic fields. You have plenty more cool stuff to discover, because this chapter is all about magnetism. I start with permanent magnets and go on to discuss the forces due to magnetic fields and what happens to charges in magnetic fields.

 Satellites that need to be oriented in real time to point to stars, the moon, or to locations on Earth are often steered by magnetic coils that push against the Earth's magnetic field instead of relying on gas jets. Magnetism is even a force in space!

Finding the Source of Attraction

If you've ever held two magnets in your hand, you know that there can be forces between them, pulling them together or pushing them apart. Those forces are the result of magnetic fields created at the microscopic level.

Usually, all the atoms in an object aren't oriented in any particular way, so all the little magnetic fields that the atoms generate cancel each other out. However, with some materials, like iron, you can orient the atoms so that a significant number of their tiny magnetic fields point in the same direction, making the object magnetic. When an object is magnetic without the application of an external current, it's called a *permanent magnet*. You can see two permanent magnets in Figure 18-1. As you can see, the magnets exert a force on each other.

Figure 18-1:
Forces exerted by permanent magnets.

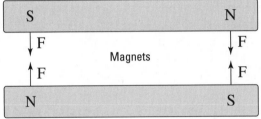

Magnetism is like electricity in that it features some positive and negative aspects, too, in the form of *magnetic poles*. Just as electric fields go from positive to negative charges, magnetic fields go from one pole to the other. The poles in magnetism are called the *north pole* and the *south pole*.

The names of the poles come from the use of permanent magnets in compasses, because the north pole orients in a northerly direction in the Earth's magnetic field.

Magnetic fields go from the north pole to the south pole, as you can see in Figure 18-2, which shows the magnetic field of a permanent bar magnet.

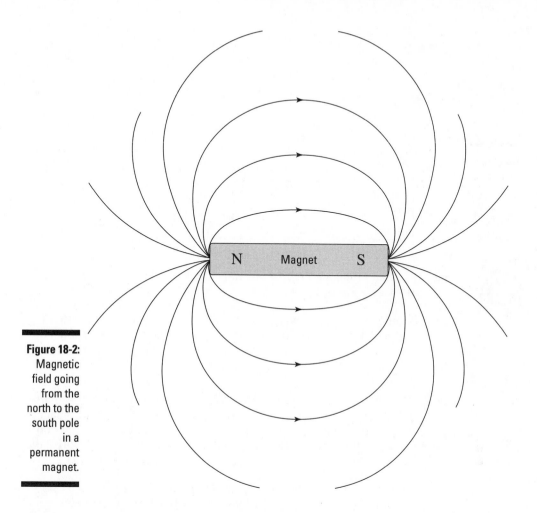

Figure 18-2:
Magnetic field going from the north to the south pole in a permanent magnet.

Forcing a Moving Charge

Magnets have an effect on electrical current — they exert a force on the charges moving in the currents. However, the charge has to be moving. A magnetic field doesn't exert a force on a motionless charge.

You can see how this works in Figure 18-3, where a moving charge has, to its surprise, started moving through a magnetic field given by the vector B in the figure. (Why does B stand for magnetic field? The only reason I can think of is that all the other letters must have been used up.) (See Chapter 4 for more on vectors.) The magnetic field creates a force on the moving charge. Which way does the force go? You can see the answer in Figure 18-3, and there's a new right-hand rule that lets you determine that answer for yourself.

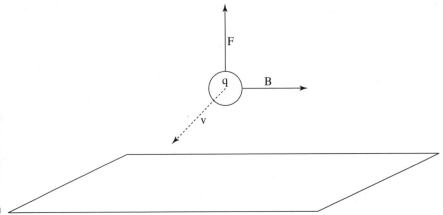

A new right-hand rule operates for moving charges, and there are two versions of it — use whichever one you find easier:

- **Version 1:** If you place the fingers of your open right hand along the magnetic field — the vector B in Figure 18-3 — and your right thumb in the direction of the charge's velocity, v, the force on a positive charge extends out of your palm. For a negative charge, reverse the direction of the force.

- **Version 2:** Place your fingers in the direction of the velocity of the charge, v, and then wrap those fingers by closing your hand through the smallest possible angle (less than 180°) until your fingers are along the direction of the magnetic field, B. Your right thumb will point in the direction of the force.

Working with these rules may remind you of working with torque (see Chapter 10). The force vector is out of the plane formed by the v and B vectors. Whichever right-hand rule you choose will work. You can now find out the direction the force acting on a moving charge. But how *big* is that force? It's time to get quantitative.

Figuring the Quantitative Size of Magnetic Forces

Knowledge of the forces of magnetism comes in handy when you're working with magnets. For example, you can determine the actual force, in Newtons, on a charged particle in motion though a magnetic field. That force turns out to be proportional to both the magnitude of the charge and the magnetic field, which makes sense.

The magnetic force is also proportional to the component of the velocity *perpendicular* to the magnetic field. In other words, if the charge is moving along parallel to the magnetic field direction, no force will act on that charge. If the charge is moving at right angles to the magnetic field, maximum force will act on the charge. Putting this information about charges together gives you the equation to find the magnitude of the force on a moving charge, q, where θ is the angle (between $0°$ and $180°$) between the v and B vectors:

$$F = qvB \sin \theta$$

Actually, this equation is a little backward. It turns out that in physics, magnetic field is actually defined in terms of the strength of the force it exerts on a positive test charge. So, here's the formal definition of magnetic field, from a physics point of view:

$$B = F / (qvB \sin \theta)$$

for a force, F, on a positive test charge, q, moving at velocity, v, such that θ is the angle (between $0°$ and $180°$) between the velocity and the field.

The units of magnetic field, in the (MKS) system (see Chapter 2), are the Tesla, T. In the (CGS) system (also see Chapter 2), the units are the gauss, G. You can relate the two units of measure like this:

$$1.0 \text{ G} = 1.0 \times 10^4 \text{ T}$$

Say, for example, that you're taking your pet electron for a walk, and a magnetic field of 12.0 Teslas appears (a huge magnetic force, given that the earth's magnetic field on its surface is about 0.6 gauss, or 6.0×10^5 T). What force acts on your pet electron if it's zipping around at a speed of 1.0×10^6 meters per second in a direction perpendicular to the field? The magnitude of the force is given by

$$F = qvB \sin \theta$$

so all you have to do is plug in the numbers:

$$F = qvB \sin \theta = (1.6 \times 10^{-19})(1.0 \times 10^6)(12.0) \sin 90° = 1.92 \times 10^{-12} \text{ N}$$

The force acting on your pet electron is 1.92×10^{-12} N, which doesn't sound like a whole lot, but remember: Pet electrons are real lightweights at 9.11×10^{-31} kg. So, what's the acceleration of the electron? Using the equation Force = mass times acceleration, you get

$$a = \frac{F}{m} = \frac{1.92 \times 10^{-12}}{9.11 \times 10^{-31}} = 2.11 \times 10^{18} \text{ m/s}^2$$

That's an acceleration of about 210,000,000,000,000,000 g, where g is the acceleration due to gravity on the Earth's surface, which is pretty stiff, even for an electron. On the other hand, if your pet electron decided to travel along the magnetic field, no force would act on it at all.

Moving in Orbits: Charged Particles in Magnetic Fields

When you have a positive charge in an electric field, like inside a parallel plate capacitor (see Chapter 17), the charge will be pushed in the opposite direction that the field lines are going, because the field lines come from the charges on the positive plate, which will repel a positive charge. The story is different when the field in question is a magnetic field, however, because there are more right angles involved. Take a look at Figure 18-4, which shows the path of a positive charge moving in a magnetic field.

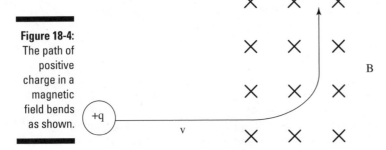

Figure 18-4:
The path of positive charge in a magnetic field bends as shown.

See all the Xs in the figure? The Xs are the ways physics shows that in this case, the magnetic field goes *into* the page. Physics intends them to be the end of vector arrows, seen tail-on (picture looking down the end of a real arrow, with the tail toward you). The positive charge travels along in a straight line until it enters the magnetic field, when a force appears on the charge. As you can verify with the right-hand rule, the force points up and will make the path of the charged particle bend, as you see in the figure.

Magnetic fields do no work . . .

Because you're a physicsmeister, you're probably asking some questions about charged particles. You know a force acts on a charged particle in a magnetic field, but how much *work* does the magnetic field do on this charge? That's a good question. When a charge moves in an electric field, the field does work on it, which introduces the concept of potential, or the work done on the charge, W, divided by the magnitude of the charge, q (work per Coulomb):

$$V = W / q$$

So, what work is done by a magnetic field on a charge? You can calculate work this way (introduced in Chapter 6):

$$W = Fs \cos \theta$$

where s is distance. Uh oh. Have you spotted the problem? θ is the angle between the force and the displacement over which that force acts. But, as you can verify with the right-hand rule, θ is always equal to 90° for charges in magnetic fields, and cos 90° = 0, which means that the work done by a magnetic field on a moving charge is zero.

Work is another case where electric fields act in a different manner than magnetic fields. An electric field most definitely does work on charges. Because a magnetic field doesn't do any work on a moving charge, it can't change that charge's kinetic energy.

. . . but they still affect moving charged particles

For all its opposition to do work on a moving charged particle, a magnetic field can, and does, change the *direction* of a moving charge's motion. In fact, a magnetic field will always change a moving charge's direction if that direction is free to change, because the force on the charge is always perpendicular to the charge's motion.

Happen to recall any other type of motion where the direction of motion is perpendicular to the force being applied? Yep, circular motion, which I feature in Chapter 7. You can see that the motion of the charge in Figure 18-4 bends when it moves through a magnetic field. Because the way magnetic fields work makes the force on the charge perpendicular to the direction of motion, you can get circular motion for charges traveling entirely inside magnetic fields.

Take a look at Figure 18-5, where a positive charge is moving to the left in a magnetic field. The B field is moving up and out of the page. How can you tell? See all those dots with circles around them? Just as an X represents the arrow of a vector seen from behind, a dot with a circle represents an arrow coming at you. So, in this case, the B field is coming at you, up and out of the page.

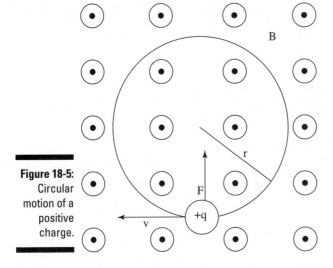

Figure 18-5:
Circular
motion of a
positive
charge.

So, the B field is coming out of the page, and the positive charge is moving to the left. Using your right hand, you can tell that the resulting force is upward (see the section "Forcing a Moving Charge" for more on the right-hand rules). The charge reacts to the upward force by moving upward. And because the force due to the magnetic field is always perpendicular to the direction of travel, the force changes direction, too. Here's the magnitude of the force:

$$F = qvB \sin \theta$$

Because v is perpendicular to B in this case, $\theta = 90°$, so $\sin \theta = 1.0$, which means you get

$$F = qvB$$

Because the force is always perpendicular to the motion, you get circular motion. In other words, the force provides the centripetal force needed to give you circular motion (see Chapter 7):

$$F = mv^2 / r$$

where m is the mass and r is the radius of the orbit. So, you get

$$qvB = mv^2 / r$$

Solving for the radius of the orbit is easy enough:

$$r = mv / qB$$

You can now solve for the radius of circular motion for a charge q of mass m traveling in a magnetic field of magnitude B at speed v. The stronger the magnetic field, the tighter the radius. And the faster the charge, and the more mass it has, the wider the radius.

Pushing and Pulling Currents

All the info about charges in this chapter up to this point is great, you may be thinking, but how often do you deal with charges in motion? A long time may have passed since you last worked with electrons moving in a vacuum, but in fact, you deal with charges in motion every day as electric currents.

Forces on currents

Take a look at the equation for the force on a moving charge:

$$F = qvB \sin \theta$$

where q is charge, v is speed, and B is magnetic field. Because you're a clever physicsmeister, you can shake up the equation, both dividing by time, t, and multiplying by it so you don't actually change the equation:

$$F = q / t \, (vt)B \sin \theta$$

Note that q / t is the charge passing a particular point in a certain amount of time, a characteristic you know by another name — electric current. And vt is just the length that the charges travel in that time, so you can rewrite the equation as

$$F = ILB \sin \theta$$

That's the force on a wire of length L carrying current I in a magnetic field of strength B, where the L is at angle θ with respect to B. For example, take a look at Figure 18-6, where a wire carrying current I is in a magnetic field B.

Because you think of currents as streams of positive charges in physics, it's easy to find the force on the wire. Say that I = 2.0 amps and B = 10 T. How much force per meter would the wire experience? Because the wire is perpendicular to the magnetic field, you have

$$F = ILB$$

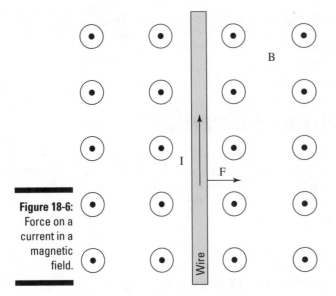

Figure 18-6:
Force on a
current in a
magnetic
field.

You know that the force per meter is

$$F/L = IB$$

so plugging in the numbers gives you

$$F/L = IB = (2.0)(10) = 20.0 \text{ N/m}$$

Twenty Newtons per meter converts to about 4.5 pounds of force per meter —
pretty hefty.

Torques on currents

Electric motors usually have permanent magnets built into them, and the
fields they create cut through electric coils that can rotate. The coils are
the objects that turn in an electric motor, which works because the force
on the electric coil generates a torque (see Chapter 11). You can see this
rotation in action in Figure 18-7.

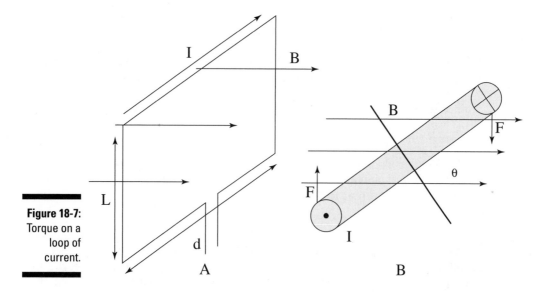

Figure 18-7:
Torque on a
loop of
current.

Take a look at diagram A in Figure 18-7. A loop of current is embedded in a magnetic field, and that field creates forces on the loop, as you can see from above
in diagram B. Those forces create two torques around the center pivot. As you
can see in diagram B, the moment arm (see Chapter 11) of each torque is

moment arm = ½ d sin θ

where *d* is the diameter of the loop. Each torque is the force — F = electric
current, I, times the force on the length of coil, L, times a magnetic field of
strength B — multiplied by the moment arm, and because there are two
torques, corresponding to the two sides of the loop, you get the total torque, τ:

τ = ILB (½ d sin θ + ½ d sin θ) = ILBd sin θ

You get an interesting result, because the product dL equals the *area* of the
coil. So, for a coil of cross-sectional area A and angle θ as shown in diagram B,
the total torque is equal to

τ = IAB sin θ

Usually, however, coils are made of multiple wires, not just a single current-
carrying loop. For example, if a coil is made up of N loops of wire, you have
to multiply the current in a single loop by N to get the total torque, which
gives you

τ = NIAB sin θ

Now you can find the total torque on a coil of N loops of wire, each carrying current I, of cross-sectional area A, in a magnetic field B, at angle θ. Whew.

A physics problem may ask you to find the maximum torque a coil of N loops will feel in a magnetic field. To find the maximum possible torque, consider the case where θ = 90°, which makes sin θ = 1.0. In this case, you get the following maximum torque

$$\tau = NIAB$$

If you have a coil with 2,000 turns, a current of 5.0 amps, a cross-sectional area of 1.0 m^2, and a magnetic field of 10.0 T, for example, what's the maximum possible torque? Easy enough:

$$\tau = NIAB = (2,000)(5.0)(1.0)(10.0) = 1.0 \times 10^5 \text{ N-m}$$

You get a maximum torque of 1.0×10^5 N-m, which is very large — an effect of having so many loops of wire. If you have only a single loop of wire, the maximum torque would be only 50 N-m, which is the reason you see so many turns of wire in the rotating parts of electric motors.

Identifying the Magnetic Field from a Wire

Moving charges feel the effects of magnetic fields (such as by changing direction; see the section "Moving in Orbits: Charged Particles in Magnetic Fields"), but they also *create* magnetic fields. Currents are made up of moving charges, so they're handy ways to create magnetic fields.

Turning an accident into a magnetic field study

As a student, I did some studies at the National Magnet Lab at Massachusetts Institute of Technology, working around some very thick cables carrying some large currents — 1,000 A and more. One day, I tripped on one of those wires and dropped a wrench next to it. When I picked up the wrench, I could feel the powerful magnetic field generated by the cable. Moving the wrench around and at different distances from the cable, I could verify the magnetic field was indeed circular. "This is neat!" I thought to myself. The professor I was working with, however, told me to knock it off and get back to work.

You can examine the creation of magnetic fields by discussing the magnetic field generated by a single wire, as shown in Figure 18-8. When you can identify the magnetic field that comes from a single wire, you're in business, because you can break down complex currents into single wires and then add the magnetic fields from each one of them.

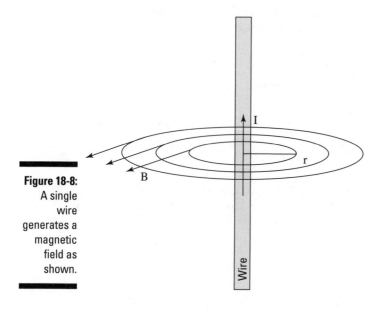

Figure 18-8:
A single
wire
generates a
magnetic
field as
shown.

The magnetic field dies off the farther you get from the wire. It decreases linearly, in inverse proportion to the distance r you are away from the center of the wire, so you have

$$B \propto 1 / r$$

The current is proportional to the current I — twice the current, twice the magnetic field. So now you have

$$B \propto I / r$$

The constant of proportionality, for historical reasons, is written as $\mu_o / 2\pi$, which means you finally get the following result for the magnetic field:

$$B = \mu_o I / (2\pi r)$$

Here, I is the current in the wire, and r is the radial distance from the center of the wire. The constant μ_o, part of the constant of proportionality, is $4\pi \times 10^{-7}$ T-m/A. This constant is called the *permeability of free space.*

Which way does the magnetic field go? You have to use another right-hand rule. If you put the thumb of your right hand in the direction of the current, your fingers will wrap around in the direction of the magnetic field. At any one point, the direction your fingers point is the direction of the magnetic field, as shown in Figure 18-8.

Say, for example, that you have a current of 1,000 A, and you're 2.0 cm from the center of a wire. How big is the magnetic field? You know that

$$B = \mu_o I / (2\pi r)$$

So, plugging in the numbers gives you

$$B = \frac{\mu_o I}{2\pi r} = \frac{(4\pi \times 10^{-7})(1000)}{2\pi (0.02)} = 0.01 \text{ T} = 100 \text{ gauss}$$

The field would be about 100 gauss.

As another example, say you have two wires parallel to each other, each carrying the same current, I, a distance r apart. What's the force on wire 1 from wire 2? You know that the force on wire 1, which is carrying current I in magnetic field B, is

$$F = ILB$$

What's B? Well, the magnetic field at wire 1 from wire 2 is

$$B = \mu_o I / (2\pi r)$$

so

$$F = \mu_o I^2 L / (2\pi r)$$

You can check with the right-hand rule to see that wires with current in the same direction will experience a force toward each other and wires with current in the opposite direction will experience a force directly away from each other.

Centering on Current Loops

"Okay," says the design team, "we need your help. Take a look at the strange setup in Figure 18-9. Ever seen anything like that before?"

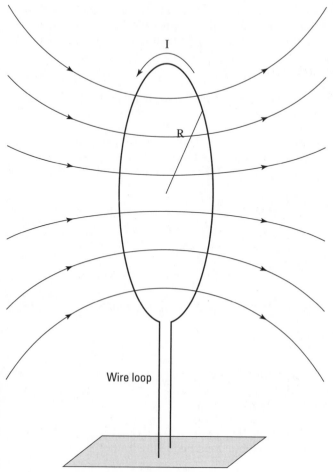

"Sure," you say, "That's a current loop."

"We know that," says the design team, "but what we want to do is to figure out the magnetic field at the very center of that loop."

"The very center?" you ask.

"The very center," they say.

"And I'll get my fee?" you ask.

"Sure thing," they say.

"Okay," you tell them — here's what the magnetic field looks like at the very center of a current loop; N is the number of turns in the loop, I is the current in the loop, and R is the radius of the loop:

$$B = N\,(\mu_o I)\,/\,(2R)$$

When you wrap the fingers of your right hand in the direction of the current, your thumb points in the direction of the generated B field.

Say, for example, that you have 2,000 turns of wire in a loop, the current is 10.0 amps, and the radius of the loop is 10.0 cm. What's the magnitude of the magnetic field in the center of the loop? Just plug in the numbers:

$$B = \frac{N\mu_o I}{2R} = \frac{(2000)(4\pi \times 10^{-7})(10.0)}{2(0.10)} = 0.126 \text{ T}$$

The magnetic field would be 0.126 T.

Achieving a Uniform Magnetic Field with Solenoids

What if you want a uniform magnetic field, just as parallel plate capacitors give you a uniform electric field (see Chapter 17)? You put a whole bunch of current loops together, as you see in Figure 18-10.

When you put multiple loops next to each other, as you see in diagram A, you get a uniform magnetic field inside the tunnel of loops (as you see in diagram B).

This new arrangement is called a *solenoid,* and it gives you a uniform magnetic field. A solenoid is just multiple loops next to each other that give you a uniform magnetic field.

What's the strength of the generated magnetic field? If the length of the solenoid is large compared to its radius, you get this equation for the magnetic field:

$$B = \mu_o nI$$

Use the right-hand rule for current loops (see the previous section) to determine the direction of the magnetic field. Here, *n* is the number of wire loops in the solenoid per meter — or the number of turns per meter — and I is the current in each turn.

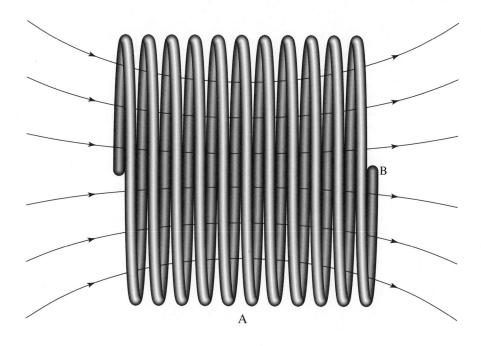

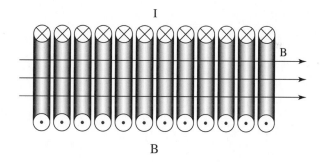

Figure 18-10:
Magnetic
field in a
solenoid.

For example, if you want to get a 1.0-T uniform magnetic field in your lab for your experiments, and you have a solenoid of 1,000 turns per centimeter, what current would you need? Just plug in the numbers:

$$I = \frac{B}{\mu_\circ n} = \frac{1.0}{\left(4\pi \times 10^{-7}\right)\left(1.0 \times 10^5\right)} = 7.96 \text{ A}$$

You need about 8 amps to get the uniform magnetic field you desire.

Chapter 19

Keeping the Current Going with Voltage

In this chapter, currents start getting wavy as you begin to study alternating current. Alternating current produces impedance, sort of a frequency-based resistance. But impedance is only one of the juicy topics coming up in this chapter. I also discuss inductance, capacitors, and inductors, all topics that that are central to physics and all things circuit-related.

Inducing EMF (Electromagnetic Frequency)

EMF, or *electromagnetic frequency*, is the rate at which energy is drawn from a source that supplies the flow of electricity in a circuit. The electricity flow is expressed in volts. If the magnetic flux (see Chapter 18) in a circuit changes for any reason, then an electric field will be generated in the circuit. This generated field is an *induced EMF*.

Magnetic flux measures the number of magnetic field lines that pass through a surface perpendicularly.

You can find the induced EMF of a conducting bar moving through a magnetic field. This voltage is induced in the bar by its motion through the magnetic field. If a circuit is present that charges can flow through, as in Figure 19-1, the charges will indeed flow, producing a current. For example, say that you have a metal bar 1.0 meter in length, moving at 60 miles per hour — about 27.0 meters per second — through a magnetic field of 1.0 T. How much voltage would appear across that bar if it travels at right angles to the magnetic field? Just plug in the values:

$$V = vBL = (27.0)(1.0)(1.0) = 27.0 \text{ V}$$

At right angles, 27.0 volts would appear across the bar. Confusing? The examples in the next two sections should help clear things up.

Moving a conductor in a magnetic field to cause voltage

Take a look at the situation you see in Figure 19-1. A conducting metal bar is moving to the right on metal rails, and the whole bar is immersed in a magnetic field going into the paper. What kind of physics is going on here?

The charges in the metal bar are moving in an electric field at speed v, so they experience a force of this magnitude:

$$F = qvB$$

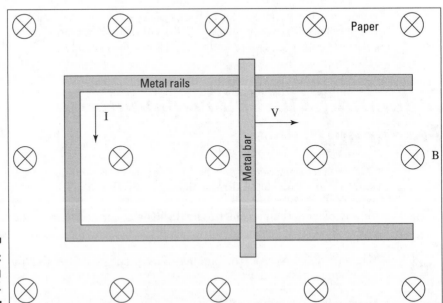

Figure 19-1:
Inducing
some EMF.

where q is electric charge and B is magnetic field. The force is enough to make the charges move, which means a *voltage* is generated, or an EMF (electromagnetic force). In other words, an electric field will be acting along the bar. The force generated on each charge in the bar due to the magnetic field will be

$$F = Eq$$

where E is the electric field. If the bar is L meters long, the electric field is equal to V / L, where V is the voltage difference between the two ends of the bar. So, you can also write the equation as

$$F = Eq = Vq / L$$

The force found in this equation is the force on the charges due to their motion in a magnetic field, so the two equations you've calculated for force are equal:

$$qvB = Vq / L$$

So, if you solve for the generated voltage from the top to the bottom of the bar, you get

$$V = vBL$$

Inducing voltage over a certain area

The change in area is another factor to consider when looking at induced EMF. If a metal bar travels a distance x in a time t through a magnetic field, for example, you can say that v = x / t, giving you this equation:

$$V = vBL = (x / t)BL$$

Now consider the product of xL — the distance the bar goes in a certain time multiplied by the length of the bar. If you take a look at Figure 19-2, you see that xL equals the *area* that the bar sweeps out as it's moving (the area appears shaded).

So, if the area is A, and the distance the bar goes in time t is x, the change in the enclosed area, ΔA, in time t is x / t, so you can write the equation for the EMF this way:

$$V = B(\Delta A / \Delta t)$$

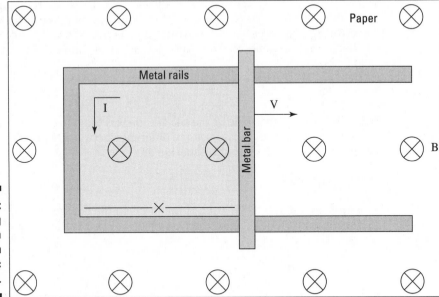

Figure 19-2:
Sweeping
out an area
with a bar in
a magnetic
field.

Factoring In the Flux with Faraday's Law

The quantity BA in the equation to find EMF — $V = B(\Delta A / \Delta t)$ — is called *magnetic flux*. Flux is a measure of how much of a magnetic field goes through a certain area — double the magnetic field, for example, and you double the flux. Magnetic flux, which has units of T-m^2 in the MKS system, is given the symbol Φ. So, the equation for EMF:

$$V = B(\Delta A / \Delta t)$$

becomes

$$V = \Delta\Phi / \Delta t$$

The equation tells you that the generated EMF is equal to the change in flux per second. But that's not the final equation. You usually see it written with a negative sign (see the next section for more information on this negative sign, which appears due to Lenz's Law):

$$V = - (\Delta\Phi / \Delta t)$$

The negative sign tells you that the generated EMF acts in such a way to create a current that will oppose the change in magnetic flux. This equation is called *Faraday's Law*. You usually see the law given in terms of magnetic flux going through a coil of N loops, as you see in Figure 19-3, where the magnetic field, B, is increasing through the coil.

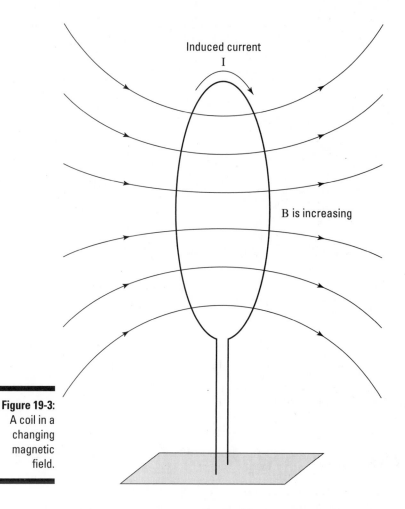

Induced current
I

B is increasing

Figure 19-3:
A coil in a
changing
magnetic
field.

If you have a coil of N loops and a changing amount of flux through that coil, Faraday's Law says the EMF induced, measured in volts, in that coil is

$$V = -N(\Delta\Phi / \Delta t)$$

How does the magnetic flux change in a coil if the coil doesn't change size? The magnetic field strength can change — as in Figure 19-3, where B is increasing — or the area that the coil presents to the magnetic field can change — as you see in Figure 19-4, which shows the coil from above.

As the angle θ changes, the amount of flux through the coil becomes

$$\Phi = \Phi_{max} \cos \theta$$

Other examples with this equation occur throughout this chapter.

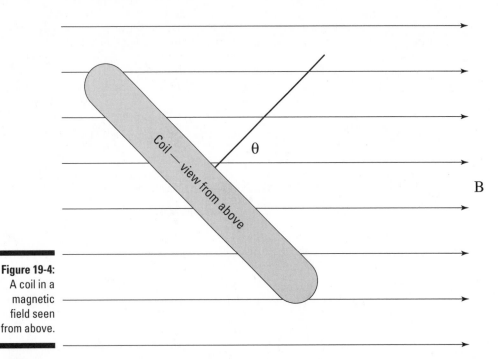

Figure 19-4:
A coil in a
magnetic
field seen
from above.

A coil has two ways to generate EMF: if the magnetic field changes or if the angle of the coil with respect to the magnetic field changes.

Getting the Signs Right with Lenz's Law

When a magnetic field starts going through a coil of wire, an EMF appears around the coil, and it takes some energy to establish that magnetic field. And when that magnetic field is established at full strength, the whole system resists any changes.

Playing around in your lab one day, say, for example, that you have a huge magnetic field going through a coil of wire, and you're taking measurements when the lights go out. However, you notice that the magnetic field through the coil doesn't disappear immediately — it diminishes slowly. Why does that happen? Because the induced EMF makes current flow in such a way that it keeps the status quo going — that is, it keeps the magnetic field unchanged.

When the lights went out, you stopped supplying the magnetic field that goes through the coil. However, the EMF exists in such a way that it tends to keep the magnetic field unchanged, although it will die away in time. That is the essence of *Lenz's Law:* An induced EMF will act so that the resulting current creates an induced magnetic field that opposes the change in flux. For example, take a look at Figure 19-3. The magnetic field being applied to the coil is increasing in time. The induced EMF will act in such a way to keep the situation as is, so it will create an *induced magnetic field* that counters the increasing magnetic field, as you can see in Figure 19-5.

Note the direction of the induced current in the coil. If you wrap the fingers of your right hand around the coil in the direction of the current, your right thumb will point in the direction of the induced magnetic field (see Chapter 18 for more on the right-hand rule). And that induced magnetic field will oppose the increasing magnetic field being applied to the coil.

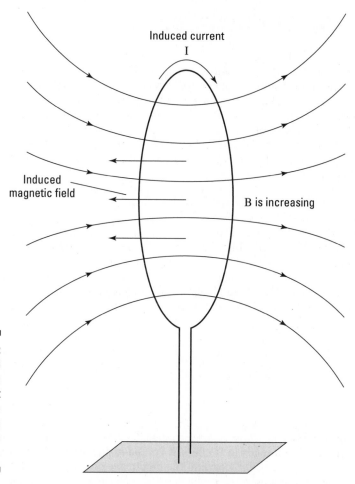

Figure 19-5:
An induced
magnetic
field that
counters an
increasing
magnetic
field.

If you keep Lenz's Law in mind, you can always figure out which way an induced current goes — it acts in a way to keep the status quo. If the magnetic flux is increasing, the induced current creates an induced magnetic field that tries to stop the flux from increasing. If the magnetic flux is decreasing, the induced current flows in such a way to increase the flux.

Test your newfound knowledge with Figure 19-5. Can you correctly predict the direction in which the current will flow, given that the flux through the coil from external sources is increasing in time? Bear in mind that the induced current will die away gradually, so it won't be able to permanently oppose the change in flux. You can see how the actual current flows shown in the figure.

Figuring out Inductance

How strong an ability does a coil have to oppose the change in magnetic flux through itself? That's a measure of the coil's *inductance.* Inductance is completely different from the concept of induced current, discussed earlier in this chapter. This being physics and all, a discussion of coils and inductance requires you to get numeric. But how can you get numeric on a coil creating an induced current? All you have to do is rewrite Faraday's Law (see the section "Factoring In the Flux: Faraday's Law") where the induced electromagnetic force is proportional to the change in current that flows through the coil. Faraday's Law says

$$V = -N(\Delta\Phi / \Delta t)$$

You can change this equation if you introduce the concept of inductance, which has the symbol L and is measured in Henries (symbol H) in the MKS system.

The total induced flux through all N loops of the coil, $N\Delta\Phi$, is proportional to the current flowing through the coil. The constant of proportionality is L. In other words,

$$N\Delta\Phi = LI$$

So, Faraday's Law becomes

$$V = -L (\Delta I / \Delta t)$$

L has become the *self-inductance,* because it's a measure of how strongly a coil will react to changes in the flux. If you have a current flowing through a coil and you change that current, the flux in the coil will change — the larger L is, the better equipped the coil will be to resist the change in the current by generating its own current. This means that a coil can resist sudden changes in the current through it because its self-inductance acts to resist any change in the flux. For this reason, coils are called *inductors* in electrical circuits.

As you find out in Chapter 16, the circuit element that works with electric fields is the capacitor. Now you know that the circuit element for magnetic fields is the inductor. Each inductor has a special quality that makes it useful in circuits. The charge in a capacitor can't change instantaneously, which means that the voltage across it can't change instantaneously. And the current through an inductor can't change instantaneously, either, which means you have to examine alternating currents.

Examining Alternating Current Circuits

The unique behavior of capacitors and resistors in circuits really comes into play when currents and voltages are changing in time — in other words, when you have *alternating current,* or *A/C.* The current that you get from a house outlet is alternating current, for example. In A/C, the direction of both the current and the voltage reverse periodically. You can see an example in Figure 19-6, which shows an A/C circuit.

The circle at the bottom with the squiggle is the symbol for an A/C voltage source; you can see the alternating voltage it creates at the top of Figure 19-6.

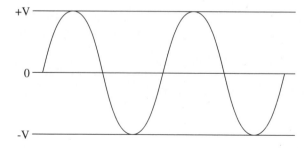

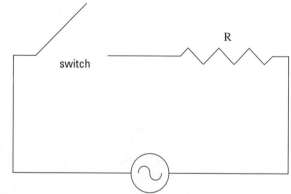

Figure 19-6: An A/C circuit, where the direction of the current and voltage reverse.

Picturing alternating voltage

What does alternating voltage look like? Circuits provide many different ways to alternate voltage, but the most common looks like the sine wave you see in Figure 19-6, which shows the voltage you get from a standard wall socket.

Borrowing some concepts from circular motion (see Chapter 7), here's how you express the voltage mathematically:

$$V = V_o \sin 2\pi f t$$

where V_o is the maximum voltage, f is the frequency with which the voltage cycles (60 Hz for wall current in the United States), and t is time. What's the current that flows as a result of that voltage when the switch is closed? The only circuit element (besides the switch, which doesn't count) is the resistor (see Chapter 17), and the resistor doesn't react to changing voltage as capacitors and inductors do (by trying to keep the voltage or the current constant). For a resistor, $V = IR$ (where I is current and R is resistance), no matter how the voltage is changing, so the current in the circuit is

$$I = (V_o / R) \sin 2\pi f t$$

What about the power dissipated in a circuit (power dissipated by the resistor as heat)? Chapter 17 tells you that Power (P) = IV, but because the current and voltage change in time, you can't say that $P = I_o V_o$. On average, in fact, you get

$$P = \tfrac{1}{2} I_o V_o$$

You often see the equation written like this:

$$P = \frac{I_o}{\sqrt{2}} \frac{V_o}{\sqrt{2}} = I_{rms} V_{rms}$$

The quantities I_{rms} and V_{rms} are called the *root mean square current* and *root mean square voltage* — the maximum current divided by the square root of two and the maximum voltage divided by the square root of two.

Unearthing root mean square current and voltage

Using the root mean square values for current and voltage, you can give the power dissipated in the resistor of a circuit by using any of the following expressions (which follow the ways of finding the dissipated power for constant currents and voltages):

$$P = I_{rms} V_{rms} = I_{rms}^2 R = (V_{rms}^2) / R$$

A resistor deals with alternating current with the equation V = IR, which is always true when V is the voltage across the resistor and I is the current through it. Therefore, if the voltage across the resistor is

$$V = V_o \sin 2\pi ft$$

that means this is the current through that resistor:

$$I = (V_o / R)\sin 2\pi ft$$

So, if the voltage across the resistor appears at the top in Figure 19-7, the current through the resistor appears at bottom in the figure.

Resistor

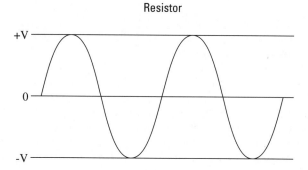

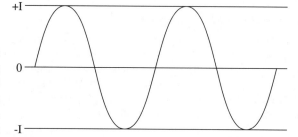

Figure 19-7:
Voltage and
current in a
resistor.

That's simple enough, but how do capacitors and inductors react to changing voltages and currents?

Leading with capacitors

Take a look at the capacitor from an A/C circuit in Figure 19-8.

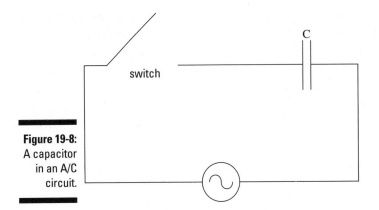

Figure 19-8:
A capacitor
in an A/C
circuit.

You find in the previous sections that $V_{rms} = I_{rms}R$ for resistors in an A/C circuit. How can you relate the voltage across a capacitor to the current flowing through it? With the following equation:

$$V_{rms} = I_{rms} X_C$$

The equation has the same form as $V_{rms} = I_{rms}R$, but what on earth is X_C? That's the *capacitive reactance* of a capacitor, which gives you an measure of how much a capacitor can actually act like a resistor when the frequency changes. You measure it in ohms, just like resistance for a resistor, and it's been measured experimentally to equal

$$X_C = 1 / (2\pi f C)$$

where f is frequency and C is capacitance.

Measuring currents through capacitors

You can figure out the magnitude of currents through capacitors given the applied voltage. For example, say the capacitor in Figure 19-8 is 1.00 μF (1.00×10^{-6}F — you measure capacitance in Farads; see Chapter 16), and the root mean square voltage of the voltage source is 12 V. What's the current that flows when the frequency is 1.0 Hz and when it's 10,000 Hz? Putting together the equations you just saw, you know that

$$I_{rms} = 2\pi f C V_{rms}$$

So, for 10 Hz, you get

$$I_{rms} = 2\pi f C V_{rms} = 2\pi(10.0)(1.00 \times 10^{-6})(12) = 7.54 \times 10^{-4} \text{ A}$$

And for 10,000 Hz, you get

$$I_{rms} = 2\pi f C V_{rms} = 2\pi(10^{4})(1.00 \times 10^{-6})(12) = 0.754 \text{ A}$$

Quite a difference, and all because the frequency changes. As you can see, you really have to keep track of the frequency when it comes to capacitors (unlike resistors).

Reviewing real values

The root mean square current and voltage let you deal with quantities that vary in time. A capacitor alternately absorbs and discharges power — none is lost to heat — so a capacitor uses no net energy in an A/C circuit (a resistor does, however). Because current and voltage vary with time, the *time-averaged value* (that's the root mean square value, as introduced in the sidebar "Sine and cosine waves" earlier in this chapter) is a good indication of what's happening, which is why the 110 volts you hear about for wall-socket voltage is so standard. But what if you don't want a time-averaged value? What if you want the real value, time dependent and all? Say, for example, that the voltage generated by the source in Figure 19-8 is as shown in Figure 19-9. What happens to the current?

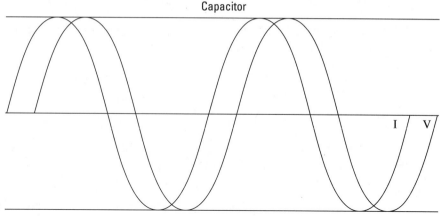

Capacitor

I / V

Figure 19-9:
Voltage and
current in a
capacitor.

As you can see in the figure, the current follows the same shape as the voltage, but it's offset a little to the left. Because the X-axis represents time, you can see that the current reaches a peak before the voltage — in physics, you say that the current *leads* the voltage. In fact, the current peaks one-quarter cycle ahead of the voltage, which means that the current gets to a particular height, such as its peak value, *before* the voltage does. So, if (note that I'm substituting ω for $2\pi f$ here, as you saw when discussing angular motion in Chapter 7)

$$V = V_o \sin(\omega t)$$

you can say that

$$I = I_o \sin(\omega t + \pi/2)$$

Here's how you can think of this relationship. Say, for example, that the current peaks and starts going down. Even after the current peaks and decreases in magnitude, it's still positive, meaning it's dumping more charge on the capacitor. And because the voltage across the capacitor is V = Q / C, where Q is charge and C is capacitance, the voltage keeps going up while the current is positive. Only when the current goes negative (crossing the X-axis) and charge is coming off the capacitor does the voltage start to decrease.

In physics terms, the fact that the current leads the voltage means that the current and the voltage are out of *phase*. In a resistor, V = IR, and there's no time dependence — the voltage and current are always in phase. In a capacitor, the current leads the voltage by a quarter cycle — which is $\pi/2$, or $90°$ — so the current and voltage are out of phase by $90°$. You can express this for a capacitor by saying that the current leads the voltage by $90°$ or that the voltage *lags* the current by $90°$.

The amplitude of the voltage and current is usually different in a capacitor — I just draw them with the same amplitude in Figure 19-8 to make the phase difference clear.

Numerically, if the following is the voltage supplied to the capacitor:

$$V = V_o \sin(\omega t)$$

you know that

$$I = I_o \sin(\omega t + \pi/2)$$

In trig, $\sin(\omega t + \pi/2) = \cos(\omega t)$, so you get

$$I = I_o \cos(\omega t)$$

This equation makes the phase difference between voltage and current in a capacitor very clear: One varies as the sine of ωt, the other varies as the cosine of ωt, and sine and cosine are out of phase by $90°$.

You can rewrite this equation for current in terms of V_o like this:

$$I = (V_o / X_C)\cos(\omega t)$$

Lagging with inductors

Take a look at Figure 19-10, which shows a squiggly thing that looks like a spring. The squiggly object is an inductor, which, like capacitors, react to alternating voltage, and the squiggly symbol is meant to look like a coil.

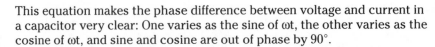

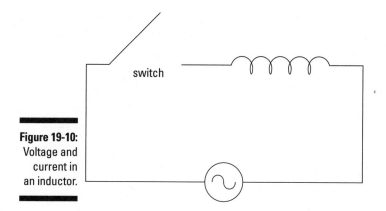

Figure 19-10:
Voltage and
current in
an inductor.

How does an inductor react to an alternating voltage placed across it? Just as with capacitance, which gives you

$$V_{rms} = I_{rms} X_C$$

You have a root mean square equation for inductors, which looks like

$$V_{rms} = I_{rms} X_L$$

What's X_L? It's the *inductive reactance* of an inductor, which is a measure of how much the inductor will resist the voltage across it, much as resistance works for a resistor. It's measured in ohms and has been measured to equal

$$X_L = 2\pi f L$$

where L is the inductance of the inductor, measured in Henries, H. The inductive reactance is directly proportional to the inductance, and the capacitive reactance is inversely proportional to the capacitance:

$$X_L \propto L$$
$$X_C \propto 1/C$$

Like a capacitor, an inductor alternately absorbs and discharges power, so it uses no net energy in an A/C circuit. If the voltage source supplies a voltage with the following time dependency:

$$V = V_o \sin(\omega t)$$

what does the current look like in the inductor? You can see the answer in Figure 19-11 — this time, the current *lags* the voltage, and the voltage *leads* the current, the reverse situation that you have with a capacitor.

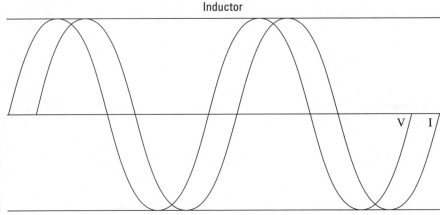

Figure 19-11:
Voltage
leads the
current in
an inductor.

Why are the current and voltage out of phase — in the opposite direction from capacitors? Take a look at the graph in Figure 19-11. When the current is at its maximum or minimum values, it isn't changing very fast. Consequently, the induced voltage, which counters any changes in the flux in the inductor, is zero at that point, which is why the current and voltage are out of phase. If the applied voltage is

$$V = V_o \sin(\omega t)$$

because the current lags the voltage, you have the following for the current:

$$I = I_o \sin(\omega t - \pi/2)$$

The current in an inductor *lags* the voltage by a quarter cycle, which means that the current gets to a particular height, such as its peak value, *after* the voltage does.

You can also write the equation as

$$I = -I_o \cos(\omega t)$$

The equation makes the phase difference between voltage and current in a capacitor very clear: One varies as the sine of ωt, the other varies as the cosine of ωt, and sine and cosine are out of phase by 90°.

You can rewrite this equation for current in terms of V_o like this:

$$I = (V_o / X_C) \cos(\omega t)$$

Handling the Triple Threat: RCL Circuits

Figure 19-12 shows you a triple threat: a resistor, a capacitor, and an inductor, all in the same circuit.

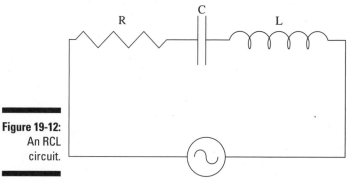

Figure 19-12:
An RCL circuit.

How the heck do you handle a circuit like this? Now you have a resistance (see Chapter 17), a capacitive reactance (see "Leading with Capacitors" in this chapter), and an inductive reactance (see "Lagging with Inductors" in this chapter), all in the same circuit. Can you just say the following?

$$V_{rms} = I_{rms}\,(R + X_C + X_L)$$

No, unfortunately, you can't, because the capacitor and inductor don't vary the same way in time, so you can't sum their voltages. Instead, you have to introduce a new quantity — *impedance* — which has the symbol Z:

$$V_{rms} = I_{rms}\,Z$$

What's Z? Here's how you express it in terms of R, X_C, and X_L:

$$Z = \sqrt{R^2 + (X_L - X_C)^2}$$

For example, say that at a particular frequency, X_L = 16.0 Ω, X_C is 12.0 Ω, R is 3.0 Ω, and V_{rms} = 10 V. What would the root mean square current be? You know that

$$Z = \sqrt{3.0^2 + (16.0 - 12.0)^2} = 5.0\ \Omega$$

so, using the equation $V_{rms} = I_{rms} Z$, you get

$$V_{rms} / Z = I = 10.0 / 5.0 = 2.0 \text{ A}$$

You get 2.0 amps.

You can also determine whether the current leads or lags by determining the angle between the current and the voltage, Φ. Here's how you determine the tangent of that angle:

$$\tan \Phi = (X_L - X_C) / R$$

Chapter 20

Shedding Some Light on Mirrors and Lenses

. . . .

In This Chapter

▶ Reviewing the basics of mirrors

▶ Observing light as it bends

▶ Examining the behavior of flat, concave, and convex mirrors

▶ Looking through converging and diverging lenses

. . . .

*T*his chapter illuminates light and what happens to it under various conditions. I discuss the behavior of light as it passes through the water (the light gets bent, as you can tell when you go fishing), goes through a lens (the light can get concentrated, as when you start paper burning with focused sunlight, or gets diverged, as in glasses for nearsighted people), and reflects off a mirror.

All about Mirrors (srorriM tuoba llA)

Light reflects from mirrors, and physics has a lot to say about how that works. Take a look at the unusual situation you see in Figure 20-1. Light is coming in from the left and bouncing off a mirror. As you can see, the light comes in at a certain angle with respect to the normal (the *normal* is the line drawn perpendicular to the mirror), called the *angle of incidence,* θ_i. And it bounces off at the *angle of reflection,* θ_r.

The *Law of Reflection* says that the angle of incidence and the angle of reflection are equal:

$$\theta_i = \theta_r$$

In other words, if the light comes in at a 30-degree angle, it'll bounce off at a 30-degree angle.

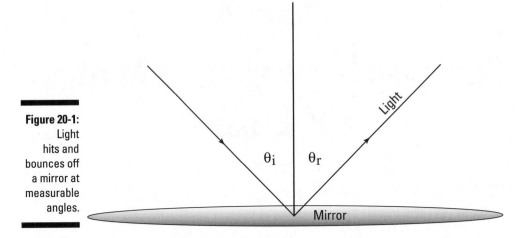

Figure 20-1:
Light
hits and
bounces off
a mirror at
measurable
angles.

When Light Gets Bendy

Take a look at the situation in Figure 20-2. Light is striking a block of glass at a certain angle, θ_1, with respect to the normal (see the previous section) and enters the glass, ending up at an angle of θ_2 with respect to the normal.

Refracting light with Snell's Law

How do you handle, physics-wise, light that bends after it enters a block of glass? You can relate the before and after angles, θ_1 and θ_2, this way:

$$n_1 \sin \theta_1 = n_2 \sin \theta_2$$

What are n_1 and n_2? They're called *indexes of refraction*. When light bends as it passes through the boundary between different substances, *refraction* takes place, and you can measure the index of refraction for various materials.

The law that describes how light bends is called Snell's Law. *Snell's Law* says that when a light ray travels from a material that has the index of refraction n_1 at angle θ_1 with respect to the normal into a material that has the index of refraction n_2, the refracted ray will be traveling at angle θ_2 with respect to the normal such that $n_1 \sin \theta_1 = n_2 \sin \theta_2$. For example, the index of refraction for air is just about the same as it is for a vacuum: 1.0. The index of refraction for glass is approximately 1.5 in most cases, so you can say that if $\theta_1 = 45°$ for the situation in Figure 20-2, you have

$$1.0 \sin 45° = 1.5 \sin \theta_2$$

or

$$\sin \theta_2 = (1.0 \sin 45°) / 1.5$$

The latter equation allows you to solve for θ_2:

$$\theta_2 = \sin^{-1}(1.0 \sin 45° / 1.5) = 28.1°$$

You find that θ_2 equals 28.1°. In other words, the light is refracted toward the normal, as you see in Figure 20-2.

Examining water at apparent depths

Take a look at Figure 20-3, which shows a spear fisherman taking aim at a fish in the water. Light reflecting off the fish hits the water/air boundary and is refracted. The fisherman assumes that the light is coming straight from the fish, so he believes the fish is really at the apparent depth shown in the figure, not at its real depth.

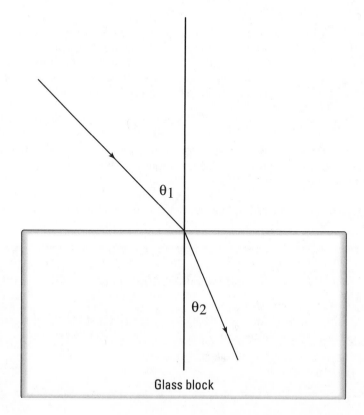

Figure 20-2:
Light entering a block of glass bends as shown.

Glass block

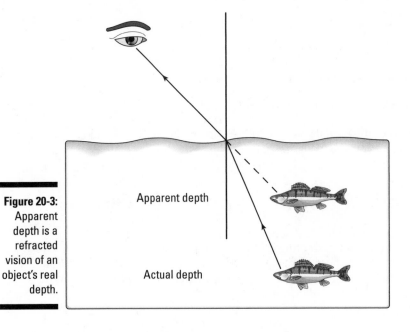

Figure 20-3:
Apparent depth is a refracted vision of an object's real depth.

Apparent depth

Actual depth

If the angles involved are small — as when the fisherman is practically over the fish — you can relate the apparent depth and the actual depth with the following equation (keep in mind that n_1 is the index of refraction of the material where the light is coming from — water in this case — and n_2 is the index of refraction of the material where the light ends up — air in this case):

Apparent depth = (Actual depth)(n_2 / n_1)

For example, if the fisherman believes the fish appears to be 2.0 meters underwater, the angles involved are small, and because the index of refraction for water is about 1.33 (and air is 1.00), you get

Actual depth = (Apparent depth)(1.33 / 1.00)

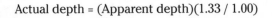

Refraction and the speed of light

The index of refraction of a material is actually the ratio of the speed of light in a vacuum divided by the speed of light in the material:

n = Speed of light in a vacuum / Speed of light in the material

So, saying that the index of refraction of glass is 1.5 is saying that light travels more slowly through glass by a factor of 1.5.

So plugging in the numbers gives you:

Actual depth = (2.00)(1.33 / 1.00) = (2.0)(1.33) = 2.66 m

The actual depth of the fish is 2.66 meters.

All Mirrors and No Smoke

You've seen any number of mirrors at work in the real world. What goes on when you look at them (or in them)? You can see a flat mirror in Figure 20-4. An object is in front of the mirror, and light from the object bounces off the mirror into your eye. From your eye's point of view, however, it looks like the light is coming from an object *behind* the mirror — a distance behind the mirror equal to the distance that the actual object is in front of the mirror. You won't find an actual object behind the mirror, however, so you say that the image is a *virtual image* of the actual object.

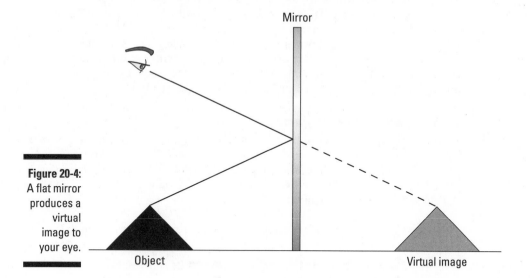

Figure 20-4:
A flat mirror produces a virtual image to your eye.

So far, so good. But what happens when mirrors start to get curvy?

Expanding with concave mirrors

When mirrors get curvy, physicists have plenty to think about. Examine the mirror in Figure 20-5. The mirror is *concave,* meaning that it looks like part of the inside of a sphere.

An easy way to identify a concave mirror is to remember that it's cupped, forming a "cave."

So, what happens when you place an object near a concave mirror?

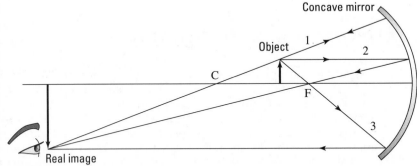

Two points are important for concave mirrors: the center of curvature, C, and the focal point, F. C is a distance equal to the radius of curvature of the sphere from which the mirror is made. F is where light coming into the mirror parallel to (and not far from) the horizontal axis is focused. For a concave mirror, F = C / 2. In Figure 20-5, you see an object between the center of curvature of the mirror. So how do you deal with that? Where will the image of this object appear? You need some more equations, your physics ammo.

Shining light on ray diagrams

You can use the three rays of light — labeled 1, 2, and 3 — you see in Figure 20-5 to find out where an object placed between the center of curvature and the focal point will appear. These three rays of light come from the object, bounce off the mirror, and come together in the image of the object. Here's how you construct these rays for a mirror:

- ✔ **Ray 1** leaves the object, bounces off the mirror, and goes through the center of curvature.

- ✔ **Ray 2** leaves horizontally from the object to the mirror, bounces off, and goes through the focal point.

- ✔ **Ray 3** travels from the object through the focal point, bounces off the mirror, and ends up going parallel to the horizontal axis.

The point where the three rays meet is the location of the image. In Figure 20-5, you can see that the image location is past the center of curvature; it's an inverted (upside-down), enlarged image. Because the image is on the same side of the mirror as the object, it's called a *real image*. A real image allows you to place a screen so the physical light from the object will focus there, creating the image.

Now take a look at a situation where the information is reversed — the object is past the center of curvature, as shown in Figure 20-6. Where does the image end up this time? You can use the same three rays, as shown in Figure 20-6. In this case, the image ends up between the center of curvature and the focal point; it's an inverted image compared to the object, reduced in size, and also a real image.

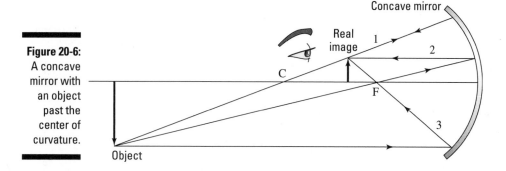

Figure 20-6:
A concave mirror with an object past the center of curvature.

Do you have to watch out for any other placement possibilities? Yes, the object can appear even closer to the mirror, between the focal point and the mirror (as you see in Figure 20-7). You can apply the same three rays here, but when you do, they don't come together in front of the mirror; they come together behind the mirror, as you see in the figure, because the image is *virtual* — no actual light comes together to form this image. Instead, when you look at the mirror from the front, the rays that bounce off the mirror seem to be coming from a virtual image behind the mirror.

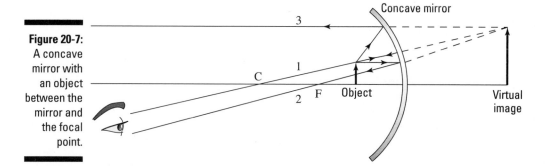

Figure 20-7:
A concave mirror with an object between the mirror and the focal point.

Crunching numbers on ray diagrams

You can calculate where an image will form with respect to a concave mirror if you have the right equation. Take a look at Figure 20-8, where you see two views of an object being reflected in a mirror.

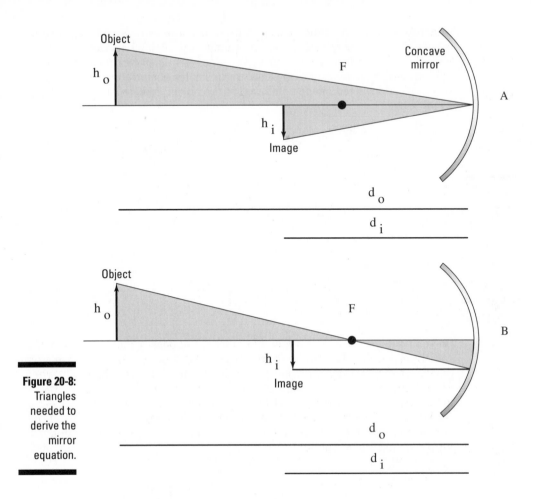

Figure 20-8:
Triangles
needed to
derive the
mirror
equation.

Here, the vital distances are

- h_o: Height of the object
- h_i: Height of the image
- d_o: Distance to the object
- d_i: Distance to the image

You can create an equation that ties all these quantities together by noticing that each of the triangles in diagrams A and B in Figure 20-8 are similar triangles, which means they have the same angles and therefore the same ratios of sides.

Diagram A is an application of the law of reflection — that the angle of incidence equals the angle of reflection — and it gives you this relation:

$$h_o / h_i = d_o / d_i$$

And from diagram B, you get

$$h_o / h_i = (d_o - f) / f$$

Setting these two equations equal to each other gives you

$$d_o / d_i = (d_o - f) / f$$

In other words,

$$1 / d_o + 1 / d_i = 1 / f$$

This equation is known as the *mirror equation;* it relates the distance the object is from the mirror and the focal length to the distance the image will form from the mirror. If the image is virtual (forming behind the mirror), d_i will be negative.

Say, for example, a cosmetics company comes to you and wants you to design a bathroom mirror that lets people see their images, only larger. And the company doesn't want the image to be upside down.

You get right on the job, deciding that Figure 20-7 will do the trick. If you place your face closer than the focal length to the mirror, you see an enlarged, upright, virtual image. Calculating swiftly, you decide that if people put their faces 12.0 cm from the mirror, you should give the mirror a focal length of 20.0 cm, which means the radius of curvature will be 40.0 cm. Where will the image form? You know that

$$1 / d_o + 1 / d_i = 1 / f$$

Plugging in the numbers gives you

$$1 / 12.0 + 1 / d_i = 1 / 20.0$$

Solving for d_i gives you –30.0 cm.

You tell the company that using a mirror with a radius of curvature of 40 centimeters will produce an image at –30.0 centimeters.

The cosmetics executives look at each other and say, "But what's the *magnification?*" "Hm," you think. "Good question."

Figuring the magnification

The *magnification* of a mirror is the ratio of the height of the image divided by the height of the object, h_o/h_i; such a magnification might be of interest to people who want to use the mirror to apply cosmetics, for example You've found that

$$h_o / h_i = d_o / d_i$$

which tells you that

$$m = -d_i / d_o$$

The negative sign is used because the magnification is positive if the image is upright and negative if the image is inverted. So, what's the magnification for the cosmetics mirror you worked on in the previous section? Well, the image forms at –30.0 cm when the object is at 12.0 cm, so the magnification is

$$m = 30.0 / 12.0 = 2.5$$

The magnification is 2.5 when someone's face is 12.0 cm from the mirror.

Contracting with convex mirrors

A *convex* mirror is likened to a mirrored section of a sphere seen from the outside. You can see a convex mirror in Figure 20-9. How do you handle this in physics?

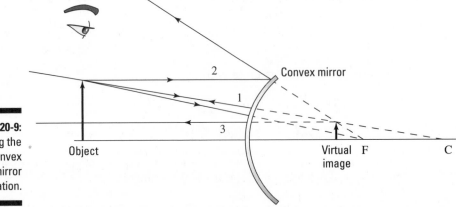

Figure 20-9:
Deriving the convex mirror equation.

No problem. You can use the three rays you use for a concave mirror in the previous section — the only difference is that the image in a convex mirror will always be virtual, and you always place the focal point and center of curvature on the other side of the mirror from the image.

Take a look at the rays in Figure 20.9; as you can see, the image created is virtual (behind the mirror), upright, and reduced. You can verify these characteristics with a metal salad bowl. If you look at yourself in the bowl from the outside, you get a smaller (and somewhat distorted) image of yourself.

You can use the mirror equation for convex mirrors as well (see the section "Crunching numbers on ray diagrams" earlier in this chapter) — if you keep in mind that because the focal point is considered to be behind the mirror, f is negative. For example, say that you have a convex mirror of focal length −20 cm, and you place an object 35 cm in front of it. Where will the image form? Just plug the numbers into the mirror equation:

$$1 / d_o + 1 / d_i = 1 / f = 1 / 35.0 + 1 / d_i = 1 / {-20}$$

Solving for d_i gives you

$$d_i = -12.7 \text{ cm}$$

The image will appear to be 12.7 cm on the other side of the mirror. Can you also find the magnification? Sure. From the previous section, you know that

$$m = -d_i / d_o = 12.7 / 35.0$$

so

$$m = 0.36$$

The virtual image will be upright and reduced by a factor of 0.36 and will appear to be 12.7 cm on the other side of the mirror. Now you've conquered mirrors, as any good physicsmeister has to.

Seeing Clearly with Lenses

Besides mirrors, the other optical elements you come across every day are lenses. *Lenses* are specially built to bend the light that goes through them, focusing the light and creating images. And like mirrors, lenses can produce real images or virtual images.

You can zoom in on two types of lenses: converging and diverging lenses. I bring both into focus in the following sections.

Expanding with converging lenses

A *converging lens* bends light toward the horizontal axis. This kind of lens can focus light from an object on one side of the lens into a real image on the other side of the lens. Want to see an insect more clearly using a magnifying glass? That's a converging lens you're using.

Relating lenses and ray diagrams

You can apply ray diagrams to lenses much like you can with mirrors. Take a look at the lens in Figure 20-10. The shape of the lens tells you that it's converging lens.

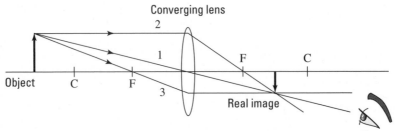

Figure 20-10: A converging lens with an object past the radius of curvature.

You can create a ray diagram by using the three rays that you see in Figure 20-10:

- ✔ **Ray 1** leaves the object and goes through the center of the lens.
- ✔ **Ray 2** leaves horizontally from the object to the lens and then goes through the focal point.
- ✔ **Ray 3** goes from the object through the focal point, through the lens, and ends up going parallel to the horizontal axis.

You can now construct the ray diagram for the case in Figure 20-10, where the object is past c, the radius of curvature. The radius of curvature is assumed to be 2f, where f is the focal length (the radius of curvature isn't exactly 2f, because lenses are really parabolic, not spherical, but it's close enough for small lenses). In this case, you get a reduced real image, which is inverted.

But what if the object is between the focal point and the radius of curvature, as you see in Figure 20-11? You can see what the three rays tell you here; you get an enlarged real image, which is also inverted.

You have one last possibility to consider: What if the image is placed closer to the lens than the focal point? You can see this illustrated in Figure 20-12. In this case, you can use rays 1 and 2, as shown in Figure 20-12, to determine that what you get is a virtual image, upright and enlarged (which is the way magnifying glasses work). As with all ray diagrams, you see where the rays come together to form an image.

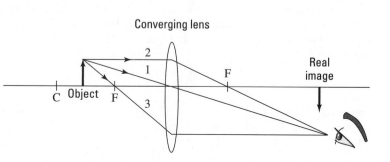

Figure 20-11:
A converging lens with an object between the radius of curvature and the focal point.

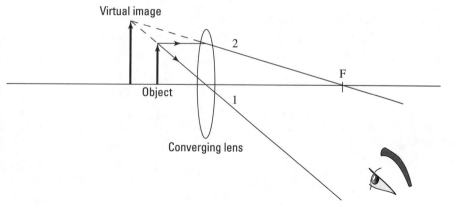

Figure 20-12:
A converging lens with an object placed closer than the focal point.

As is typical with virtual images, no actual light rays converge where the virtual image is. If you put a screen there, you wouldn't see an image. The image is virtual, which means you have to look though the lens to see it — you see the apparent image through the lens.

Bringing the lens equation into focus

How can you calculate the actual location of an image through a lens? You can use the *thin-lens equation,* which is the same as the mirror equation and can be derived in the same way:

$$1 / d_o + 1 / d_i = 1 / f$$

where d_i is the distance of the image from the lens, d_o is the distance of the object, and f is the focal length.

This equation holds for *thin* lenses (otherwise, the shape of the lens, which is assumed to be a sphere for this equation, gives you what's called *spherical aberration;* in practice, lenses are slightly parabolic). Note that if the image is virtual, d_i will be negative.

For example, if someone hands you a converging lens of focal length 5.0 centimeters to use as a magnifying glass on a postage stamp 3.0 cm away from the lens, where will the virtual image appear? All you have to do is plug in the numbers:

$$1 / d_o + 1 / d_i = 1 / f = 1 / 3.0 + 1 / d_i = 1 / 5.0$$

Solving for d_i gives you $d_i = -7.5$ cm. A negative image distance means a virtual image, and as you can see in Figure 20-12, the image will be upright and enlarged.

Because you're working with a magnifying glass, you also have to find the actual magnification.

Calculating magnification

Finding the magnification of converging lenses is the easy part. As with mirrors, you can calculate the magnification this way:

$$m = -d_i / d_o$$

If the magnification is negative, the image is inverted with respect to the object. If the magnification is positive, the image is upright compared to the object.

What's the magnification of the lens I introduce in the example from the previous section? You know that $d_o = 3.0$ cm and $d_i = -7.5$ cm, so

$$m = -d_i / d_o = 7.5 / 3.0 = 2.5$$

The magnification you get for the stamp under the magnifying glass at 3.0 cm away from the lens is 2.5.

Here's another example. Say that you want a real, enlarged image, one you can put onto a screen. For example, you may have a slide projector with an illuminated slide that you want to project onto a screen 1.0 meter away. In this case, you should use the setup you see in Figure 20-11, where the object is placed at a location between the radius of curvature (which equals 2*f*) and the focal point (*f*).

Assume that the slide is at 10 cm from the lens — what focal length should the lens have? You know that the thin lens equation tells you

$$1 / d_o + 1 / d_i = 1 / f$$

Plugging in the values gives you

$$1 / d_o + 1 / d_i = 1 / 0.10 + 1 / 1.00 = 1 / f$$

In other words,

$$10.0 + 1.00 = 11.0 = 1 / f$$

So, the focal length of the lens you need is 1 / 11.0 meters, or about 9.09 centimeters.

Contracting with diverging lenses

Converging lenses focus light toward the horizontal axis; *diverging lenses* focus light away from the horizontal axis. You can see what a diverging lens looks like in Figure 20-13. In this case, you can see that the ray diagram is giving you a virtual image, upright and reduced.

A diverging lens always produces a virtual image, upright and reduced in size compared to the object.

Figure 20-13: Using a diverging lens to create a virtual image.

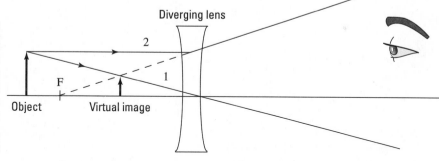

Can you apply the thin lens equation here, as you did with converging lenses? Sure can, except that you have to realize that for a diverging lens, the focal length is negative — just like a convex mirror.

For example, say that you have a diverging lens with focal length of –10.0 cm, and you place an object 4.0 cm from it. Where will the image form? Use your pal the thin lens equation for this one:

$$1 / d_o + 1 / d_i = 1 / f$$

Putting in your data gives you

$$1 / 4.0 + 1 / d_i = -1 / 10.0$$

After a little rearrangement to put d_i on one side, you get

$$d_i = -2.85 \text{ cm}$$

Note that this image forms closer to the lens than the focal length, as shown in Figure 20-13, and that the distance to the image is negative, which means that the image is virtual. What's the magnification of this lens? Using the same magnification equation you use for converging lenses (see the previous section),

$$m = -d_i / d_o$$

Plugging in numbers gives you

$$m = 2.85 / 4.0 = 0.713$$

The positive magnification tells you that the image is upright compared to the object, and because the magnification is less than 1.00, you know that the image will be reduced in size compared to the object.

Part VI
The Part of Tens

The 5th Wave   By Rich Tennant

PYTHAGOREAN GOLF

√16!

In this part . . .

1 let physics off the leash in Part VI, and it goes wild. You discover 10 Einsteinian topics, like time dilation, length contraction, and $E = mc^2$. You also take a look at 10 far-out physics facts spanning from the Earth to the far reaches of space — everything from black holes and the Big Bang to wormholes.

Chapter 21

Ten Amazing Insights on Relativity

This chapter contains 10 amazing physics facts about Einstein's Theory of Special Relativity. Well, sort of. The pieces of info I include aren't really "facts," because as with everything in physics, the information may yet be disproved someday. But the Theory of Special Relativity has been tested in thousands of ways, and so far it has been on the money. The theory gives you many spectacular insights, such as the one that states that matter and energy can be converted into each other, as given by perhaps the most famous of physics equations:

$$E = mc^2$$

You also find out that time dilates and that length shrinks near the speed of light. It's all coming up in this chapter. After you read what Albert Einstein has to say, time and space will never be the same.

Nature Doesn't Play Favorites

Einstein stated long ago that the laws of physics are the same in every inertial reference frame. In an inertial reference frame, if the net force on an object is zero, the object either remains at rest or moves with a constant speed. In other words, an *inertial reference frame* is a reference frame with zero acceleration. Newton's law of inertia (a body at rest stays at rest, and a body in constant motion stays in constant motion) applies.

Two examples of noninertial reference frames are spinning frames that have a net centripetal acceleration or otherwise accelerating frames.

What Einstein basically said is that any inertial reference frame is as good as any other when it comes to the laws of physics — nature doesn't play favorites between reference frames. For example, you may be doing a set of physics experiments when your cousin rolls by on a railroad car, also doing a set of experiments, as you see in Figure 21-1.

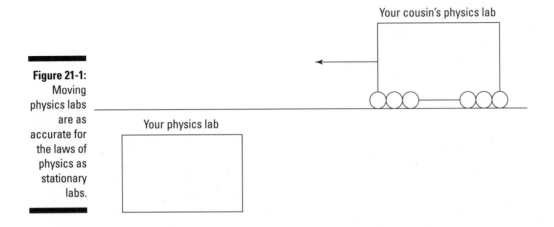

Figure 21-1:
Moving physics labs are as accurate for the laws of physics as stationary labs.

Your cousin's physics lab

Your physics lab

Neither of you will see any difference in the laws of physics. No experiment allows you to distinguish between an inertial reference frame that's at rest and one that's moving.

The Speed of Light Is Constant, No Matter How Fast You Go

Comparing speeds while you're in motion is hard enough to do with cars on the highway, let alone comparing speeds that include the speed of light. For most people, finding out that the speed of light is constant, no matter how fast you go, is unexpected. Say, for example, that your cousin, who's riding on a railroad car, finishes a drink and thoughtlessly throws the empty can overboard in your direction. It may not be traveling fast with respect to your cousin — say, 5 mph — but if your cousin's inertial reference frame (see the previous section) is moving with respect to you at a speed of 60 mph, the can will hit you with that speed added on — 65 mph. Ouch. But light will always hit you at about 670,616,629 mph.

Time Dilates at High Speeds

Imagine that you're looking up at a starry night as an astronaut hurtles past in a rocket, as you see in Figure 21-2. You think you can faintly hear the words, "Get me down, I don't care about the World Series!" coming from the rocket, but you're not sure. Einstein's Special Theory of Relativity says that the time you measure for events occurring on the rocket ship happen more slowly than events measured by the astronaut. In other words, time dilates, or "expands," from your point of view on the Earth.

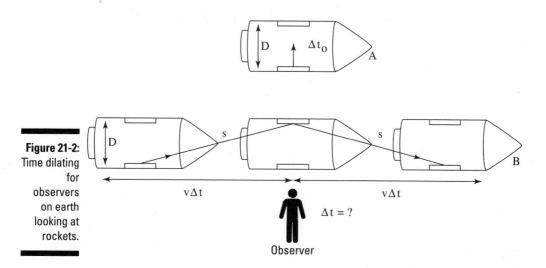

Figure 21-2:
Time dilating
for
observers
on earth
looking at
rockets.

To see how this works, take a look at Figure 21-2, diagram A. In that diagram, a special clock bounces light back and forth between two mirrors mounted on the inside walls of the rocket at distance D apart. The astronaut can measure time intervals based on how long it takes the light to bounce back and forth. From your perspective, however, the time appears different. You see the rocket ship hurtling along, so the light doesn't just have to travel the distance s — it also has to take into account the distance the rocket travels horizontally.

Space Travel Ages You Less

Don't go telling this to your beauty-obsessed, wealthy aunt, but if you're traveling in space, you may age less than someone on the ground. For example, say that you observe an astronaut who's moving at a speed of $0.99c$, where c equals the speed of light. For the astronaut, tics on the clock last, say, 1.0 second. For each second that passes on the rocket as measured by the astronaut, you measure 7.09 seconds.

This effect even takes place at smaller velocities, such as when a friend boards a jet and takes off at about 520 miles per hour. The plane's speed is so slow compared to the speed of light, however, that relativistic effects aren't really noticeable — it would take about 100,000 years of jet travel to create a time difference of 1 second between you and your friend's watches. However, physicists conducted this experiment with jets and super-sensitive, cesium-based atomic clocks, capable of measuring time differences down to 1.0×10^{-9} seconds. And the results agreed with the Theory of Special Relativity.

Length Contracts at High Speeds

Not only can you measure time for events on spaceships as longer (see the previous section), but also the very length of rockets. The length of the rocket an astronaut is on is different according to his measurements than to your measurements taken from Earth. Take a look at the situation in Figure 21-3 to see how this works.

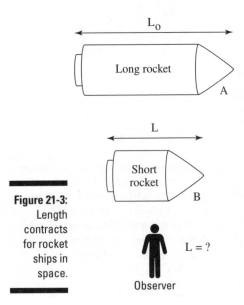

Figure 21-3: Length contracts for rocket ships in space.

The length, L_o, of an object measured by a person at rest with respect to that object will be measured as L, a shorter length, by a person moving at velocity v with respect to that object. In other words, the object *shrinks*.

Note that shrinking only takes place in the direction of motion. As you can see in Figure 21-3, the rocket ship appears to contract in the direction of motion when you measure it (diagram B), but not from the astronaut's point of view.

$E = mc^2$: The Equivalence of Matter and Energy

Einstein's most famous contribution is the about the equivalence of matter and energy — that a loss or gain in mass can also be considered a loss or gain in energy. (I'm going to resist putting in that tired physics joke about how for a dieter, a loss of mass results in an increase in energy.) What was Einstein's result? $E = mc^2$? Actually, no. It was

$$E = \frac{mc^2}{\sqrt{1 - \dfrac{v^2}{c^2}}}$$

As a special case, the object you're converting to energy may be at rest, which means that $v = 0$. In such a case, and only in that case, you do indeed find that $E = mc^2$.

You've seen Einstein's famous equation before, but what does it really mean? In a way, you can almost think of mass as "condensed" energy, and this formula gives you the conversion factor between kilograms and Joules, which is c^2, the speed of light squared.

Matter Plus Antimatter Equals Boom

You can get a complete conversion of mass into energy when you have both matter and antimatter. *Antimatter* is just like standard matter but sort of reversed. Rather than electrons in the atoms of antimatter, you have positively charged *positrons*. And in place of positively charged protons, you have negatively charged *antiprotons*. Science fiction aficionados may recognize antimatter as the driving force in the Starship Enterprise's engines in *Star Trek*.

But the weird thing is that antimatter actually *exists*. Scientists can locate it in the universe, and in fact, the sun is making it all the time. When a standard hydrogen atom (electron and proton) and an antimatter hydrogen atom (positron and antiproton) come together, they both get converted entirely, 100 percent, into energy. What happens to that energy? It streaks out as high-powered photons (which, please note, can transfer heat as radiant energy).

Don't make plans for the end of the sun

The sun radiates power at 3.92×10^{26} Watts. So, in one second, it radiates 3.92×10^{26} Joules of energy. This translates into the sun losing about *4.36 billion kilograms of mass per second* — whoa! That's about 4.79 million tons of matter lost every second. You can almost hear solar scientists going nuts. Surely, at that rate, there will be no more sun left in a few weeks. They cry out, "Our mortgages! We'll have to get new jobs when the sun goes out!" What they shouldn't forget is that the sun has a mass of 1.99×10^{30} kg. Even at 4.36×10^9 kg of mass lost per second, it will still last for a while. How long? If radiating away mass were the only physical mechanism at work, the sun would last $1.99 \times 10^{30} / 4.36 \times 10^9$ = 4.56×10^{20} seconds. That's about 1.44×10^{13} years, or 144 billion centuries.

The Sun Is Radiating Away Mass

Most of the energy we get from the sun comes from *fusion,* the combination of atomic nuclei into other nuclei. The sun is radiating away a heck of a lot of light every second, and for that reason, it's actually losing mass. Although the sun is losing mass as it gets converted into radiant energy, however, you have nothing to worry about; there's plenty more where that came from.

The Speed of Light Is the Ultimate Speed

Not counting *Star Trek* and other science fiction shows and books, you can't go faster than the speed of light, which is why c is the same in all inertial reference frames (see the first section of this chapter), even if the light you see is coming from an inertial reference frame coming at you. Here's what the Special Theory of Relativity says about the total energy of an object:

$$E = \frac{mc^2}{\sqrt{1 - \frac{v^2}{c^2}}}$$

For an object at rest, $E_{rest} = mc^2$. So, the relativistic kinetic energy of an object of mass *m* must be

$$KE = o = mc^2 \left(\frac{1}{\sqrt{1 - \frac{v^2}{c^2}}} - 1 \right)$$

Note that as the velocity of the object gets larger and larger, the term in parentheses above gets bigger and bigger, moving toward the infinite. So, as the velocity of the object gets infinitesimally close to *c*, the kinetic energy of the object becomes nearly infinite. And this is where, sadly, the science fiction part breaks down. As the velocity gets closer to *c*, you get closer to an infinite kinetic energy. Although that sounds impressive for rocket ships, what it really means is that you can't do it — at least not according to the Special Theory of Relativity.

Newton Is Still Right

After all the discussion about Einstein, where have physicists left Newton? What about the good old equations for momentum and kinetic energy? These equations are still right, but only at lower speeds. For example, take a look at the relativistic equation for momentum (see Chapter 9 for more on momentum):

$$p = \frac{mv}{\sqrt{1 - \dfrac{v^2}{c^2}}}$$

where *p* is momentum, *m* is mass, and *v* is velocity. Notice this part:

$$\frac{1}{\sqrt{1 - \dfrac{v^2}{c^2}}}$$

You only see a difference when you start getting near the speed of light — this factor only changes things by about 1 percent when you get up to speeds of about 4.2×10^7 meters per second, which would have been pretty big for Newton's day. At lower speeds, you can neglect the relativistic factor to get

$$p = mv$$

Newton would be happy with this result.

How about the equation for kinetic energy (see Chapter 8)? Here's how it looks in relativistic terms:

$$KE = mc^2 \left(\frac{1}{\sqrt{1 - \dfrac{v^2}{c^2}}} - 1 \right)$$

where KE is kinetic energy. Take a look at this term:

$$\frac{1}{\sqrt{1 - \frac{v^2}{c^2}}}$$

You can expand this by using the binomial theorem (from Algebra class) this way:

$$\frac{1}{\sqrt{1 - \frac{v^2}{c^2}}} = 1 + \frac{1v^2}{2c^2} + \frac{3}{8}\left(\frac{v^2}{c^2}\right)^2 + \ldots$$

When the term $v^2 \div c^2$ is much less than 1, this breaks down to

$$\frac{1}{\sqrt{1 - \frac{v^2}{c^2}}} = 1 + \frac{1v^2}{2c^2}$$

Putting that into the equation for relativistic kinetic energy gives you — guess what? Your old favorite, the nonrelativistic version (see Chapter 8):

$$KE = \tfrac{1}{2}mv^2$$

As you can see, Newton isn't left in the dust when discussing relativity. Newtonian mechanics still apply, as long as the speeds involved are significantly less than the speed of light, c (you start seeing relativistic effects at about 14 percent of the speed of light, which is probably why Isaac Newton, in the horse-and-buggy days, never noticed them).

Chapter 22

Ten Wild Physics Theories

This chapter gives you ten outside-the-box physics facts that you may not hear or read about in a classroom. As with anything in physics, however, you shouldn't really consider these "facts" as actual facts — they're just the current state of many theories. And in this chapter, some of the theories get pretty wild, so don't be surprised to see them superceded in the coming years.

You Can Measure a Smallest Distance

Physics now has a theory that there's a smallest distance, the Planck length, named after the physicist Max Planck. The length is the smallest division that, theoretically, you can divide space into. However, the Planck length — about 1.6×10^{-35} m, or about 10^{-20} times the approximate size of a proton — is really the smallest amount of length with any physical significance, given our current understanding of the universe. When you measure smaller than this length, you need quantum physics, and that means that any precise measurements are impossible. Therefore, the Planck length is the smallest length that makes physical sense with our current view of physics.

Is the Planck length *really* the smallest possible distance?

Many scholars have maintained that the Planck length is where space breaks down and that you can't get any distances smaller. Is this the truth? The Planck length basically states that when you get to smaller distances, quantum effects take over. Such effects aren't susceptible to measurement, only predictions based on probability. So, is the Planck length really the smallest possible length? Or are physicists trying to impose rules on the universe just because we don't know how forces work at those small distances? In other words — is the Planck length the smallest possible distance that we can *explain,* or is it the smallest possible distance that can *exist?* Perhaps you can find the answer in your physics explorations.

There Might Be a Smallest Time

In the same sense that Planck length is the smallest distance (see the previous section), Planck time is the smallest amount of time. The *Planck time* is the time it takes for light to travel 1 Planck length, or 1.6×10^{-35} m. If the speed of light is the fastest possible speed, you can easily make a case that the shortest time you can measure is the Planck length divided by the speed of light. The Planck length is very small, and the speed of light is very fast, which gives you a very, very short time for the Planck time:

Planck time = 1.6×10^{-35} m / $(3.0 \times 10^8$ m/s$)$ = 5.3×10^{-44} seconds

The Planck time is about 5.3×10^{-44} seconds, and times smaller than this have no real meaning as physicists understand the laws of physics right now.

Some people say that time is broken up into quanta of time, called *chronons,* and that each chronon is a Planck time in duration.

Heisenberg Says You Can't Be Certain

You may have heard of the uncertainty principle, but you may not have known that a physicist named Heisenberg first suggested it. Of course, that explains why it's called the *Heisenberg Uncertainty Principle,* I suppose. The principle comes from the wave nature of matter, as suggested by Louis de Broglie. Matter is made up of particles, like electrons. But small particles also act like waves when they travel at fast speeds, much like light waves. (Whoa, you're getting into some deep water here.)

Because particles in motion act as though they're made up of waves, the more precisely you measure their momentum, the less precisely you know where they are. You can also say the more precisely you measure their locations, the less precisely you know their locations.

Black Holes Don't Let Light Out

Black holes are created when particularly massive stars use up all their fuel and collapse inwardly to form super-dense objects, much smaller than the original stars. Only very large stars end up as black holes. Smaller stars don't collapse that far; they often end up as neutron stars instead. A *neutron star* occurs when all the electrons, protons, and neutrons have been smashed together by gravity, creating, effectively, a single mass of neutrons at the density of an atomic nucleus.

Black holes go even further than that. They collapse so far that not even *light* can escape their intense gravitational pulls. How's that? The photons that make up light aren't supposed to have any mass. How can they possibly be trapped in a black hole?

Photons are indeed affected by gravity, a fact predicted by Einstein's General Theory of Relativity (a much larger body of work than the Special Theory of Relativity (see Chapter 21), and one that took Einstein eight years to complete). Tests have experimentally confirmed that light passing next to heavy objects in the universe is bent by their gravitational fields. Gravity affects photons, and the gravitational pull of a black hole is so strong that they can't escape it.

Gravity Curves Space

Isaac Newton gave physicists a great theory of gravitation, and from him comes the famous equation

$$F = Gm_1m_2/r^2$$

where F represents force, G represents the universal gravitational constant, m_1 represents one mass, m_2 represents another mass, and *r* represents the distance between the masses. Newton was able to show that what made an apple fall also made the planets orbit. But Newton had one problem he could never figure out: how gravity could operate instantaneously at a distance.

Enter Einstein, who created the modern take on this problem. Instead of thinking in terms of gravity being a simple force, Einstein suggested in his General Theory of Relativity that gravity actually *curves* space. In other words, gravity is one of the influences that actually define what we think of as "space."

Einstein's idea is that gravity curves space (and ultimately, that's where the idea of wormholes in space comes from). In fact, to be more true to the General Theory of Relativity, you should say that gravity curves space and time. Mathematically, you treat time as the fourth dimension when working with relativity — not as a fourth spatial dimension. The vectors (see Chapter 4) you use have four components: three for the X-, Y-, and Z-axes, and one for time, *t*.

What's really happening when a planet orbits the sun is that it's simply following the shortest path through the curved spacetime through which it travels. The sun curves the spacetime around it, and the planets follow that curvature.

Matter and Antimatter Destroy Each Other

One of the coolest things about high-energy physics, also called particle physics, is the discovery of antimatter. *Antimatter* (see Chapter 21) is sort of the reverse of matter. The counterparts of electrons are called positrons (positively charged), and the counterparts of protons are antiprotons (negatively charged). Even neutrons have an antiparticle: antineutrons.

In physics terms, matter is sort of on the plus side and antimatter sort of on the negative side. When the two come together, they destroy each other, leaving pure energy — light waves of great energy, called *gamma waves.* And like any other radiant energy, gamma waves can be considered heat energy, so if you have a pound of matter and a pound of antimatter coming together, you'll have quite a bang.

As I discuss in Chapter 21, that bang, pound for pound, is much stronger than a standard atomic bomb, where only 0.7 percent of the fissile material is turned into energy. When matter hits antimatter, 100 percent is turned into energy.

If antimatter is the opposite of matter, shouldn't the universe have as much antimatter as it does matter? That's a puzzler, and the debate is continuing. Where's all the antimatter? The jury is still out. Some scientists say that there could be vast amounts of antimatter around that people just don't know about. Immense antimatter clouds could be scattered throughout the galaxy, for example. Others say that the way the universe treats matter and antimatter is a little different — but different enough so that the matter we know of in the universe can survive.

Supernovas Are the Most Powerful Explosions

What's the most energetic action that can happen anywhere, throughout the entire universe? What event releases the most energy? What's the all-time champ when it comes to explosions? Your not-so-friendly neighborhood supernova. A *supernova* occurs when a star explodes. The star's fuel is used up, and its structure is no longer supported by an internal release of energy. At that point, it collapses in on itself. But not just any star can supernova. Normally, if a star explodes, it creates a nova. But once in a while, a star supernovas, creating a practically unimaginable star explosion. Among the 100 billion stars in our galaxy, for example, the last known supernova occurred nearly 400 years ago. (I say known because light takes quite a while to reach Earth; a star could have gone supernova 100 years ago, but if it's far enough from Earth, we wouldn't know it yet.)

Most of the star that supernovas is exploded at speeds of about 10,000,000 meters per second, or about 22,300,000 miles per hour. By comparison, even the highest of explosives on Earth detonate at speeds of 1,000 to 10,000 meters per second.

The Universe Starts with the Big Bang and Ends with the Gnab Gib

Many physicists believe that the universe started, physically speaking, with the Big Bang. The idea is that the physical world all began at one location with an unimaginably huge explosion about 13.7 billion years ago. Physicists have theorized about what went on after the Big Bang, but it isn't so easy to theorize what went on before it. In fact, you can only go back to one Planck time after the Big Bang (see the section "You Can Clock a Smallest Time" earlier in this chapter for more on Planck time), because at times smaller than that, standard physics theories, including Einstein's General Theory of Relativity, break down.

Microwave Ovens Are Hot Physics

You can find plenty of physics going on in microwaves — everyday items you may have taken for granted in your pre-physics life. What really happens in a microwave oven? A device called a *magnetron* generates light waves in the part of the light spectrum called microwaves, and their wavelength is close to the size of water molecules.

When the microwaves hit the water molecules in food, they *polarize* them. In other words, the electrical field in light alternates back and forth, and microwaves have just about the right wavelength and frequency to charge one side of a water molecule and then the other, making that molecule oscillate like mad. The oscillation creates heat through friction, the oscillations of water molecules are heating your food. (For more info on electrical fields, see Chapter 16.)

Microwave ovens were invented by accident, during the early days of radar. A man named Percy Spencer put his chocolate bar in the wrong place — near a magnetron used to create radar waves — and it melted. "Aha," thought Percy. "This could be useful." And before he knew it, he had invented not only microwave ovens, but also microwave popcorn (no kidding).

Physicists May Not Have Physical Absolute Measures

Perhaps the most profound "discovery" to come out of physics is that physicists have found no true physical absolute measures.

Long ago, people saw everything in terms of absolutes. Space was fixed in place. The sun and stars rotated around the Earth. Light traveled through the aluminiferous ether, which stayed stationary. However, the study of physics has debunked all these beliefs, at least as far as many years of observation go. No fixed frame of reference has been found. No aluminiferous ether, either. Not even a fixed universal time frame. All the measurements you make are with respect to some measuring stick or inertial frame of reference that you yourself define.

Although absence of proof isn't proof of absence, the fact that physicists have found no physical absolutes could be the most important contribution that physics makes to our understanding of the universe. Perhaps, ultimately, the lesson of physics is that you're so much a part of the universe that you don't need external physical absolutes and absolute physical measures to get your bearings. You decide your own measures. And in that sense, you're already at home in the universe in which you live.

Glossary

· ·

*H*ere's a glossary of common physics terms you come across in this book:

acceleration: The rate of change of velocity per second, expressed as a vector.

adiabatic: Without releasing heat into, or absorbing heat from, the environment.

alternating current: Electrical current that changes the direction of flow periodically in time.

alternating voltage: Electrical potential that changes polarity periodically in time.

ampere: The MKS unit of measurement of current — one Coulomb per second.

angular acceleration: The rate of change of angular velocity per second, expressed as a vector.

angular displacement: The angle between the initial and final angular positions after a specified time.

angular momentum: The angular momentum of an object around a certain point is the product of the distance it is from that point and its momentum measured with respect to the point.

angular velocity: The rate of change of angular displacement per second.

Avogadro's Number: The number of items in a mole, 6.022×10^{23}.

blackbody: A body that absorbs all radiation and radiates it all away.

Boltzmann's constant: A thermodynamic constant with a value of 1.38×10^{-23} Joules per Kelvin.

capacitance: How much charge a capacitor can store per volt.

capacitor: An electric component that stores charge.

centripetal acceleration: The acceleration of an object, directed toward the center of the circle, needed to keep that object in circular motion.

centripetal force: The force, directed toward the center of the circle, that keeps an object going in circular motion.

CGS system: The measurement system that uses centimeters, grams, and seconds.

concave mirror: A mirror that has an inwardly curved surface.

conduction: The transmission of heat or electricity through a material.

conservation of energy: The law of physics that says the energy of a closed system doesn't change unless external influences act on the system.

convection: A mechanism for transporting heat through the motion of a heated gas or liquid.

converging lens: A lens that makes light rays come together.

convex mirror: A mirror that has an outwardly curved surface.

Coulomb: The MKS unit of charge.

current: The measurement of electric charge flowing per second.

density: A quantity divided by a length, area, or volume, such as mass per length, mass per volume, charge per area, or charge per volume.

direct current: The kind of electrical current that moves in one direction, such as current induced by a battery.

displacement: The change in an object's position.

diverging lens: A lens that causes light rays to spread apart.

elastic collision: A collision where kinetic energy is conserved.

electric field: The force on a positive test charge per Coulomb due to other electrical charges.

electromotive force: In a circuit, the potential difference that causes charges to flow.

electrostatic potential: The energy per positive charge needed to move a charge from one point to another.

emissivity: A property of a substance showing how well it radiates.

energy: The ability of a system to do work.

farad: The MKS unit of capacitance.

FPS system: The system of measurement that uses feet, pounds, and seconds.

frequency: The number of cycles of a periodic occurrence per second.

heat capacity: The amount of heat needed to raise the temperature of an object by one degree.

henry: The MKS unit of measurement of inductance.

hertz: The MKS unit of measurement of frequency — one cycle per second.

impulse: The product of the amount of force on an object and the time during which the force is applied.

index of refraction: A measure of the speed of light in a vacuum divided by the speed of light in a substance.

inelastic collision: A collision in which kinetic energy isn't conserved.

inertial frame: A frame of reference that isn't undergoing acceleration.

isobaric: At constant pressure.

isochoric: At constant volume.

isothermal: At constant temperature.

Joule: The MKS unit of energy — one Newton-meter.

Kelvin: The MKS unit of temperature, starting at absolute zero.

kilogram: The MKS unit of mass.

kinetic energy: The energy of an object due to its motion.

kinetic friction: Friction that resists motion of an object that's already in motion.

kinematics: Branch of mechanics concerned with motion without reference to force or mass.

latent heat: The heat needed to cause a change in phase per kilogram.

law of conservation of momentum: A law stating that the momentum of a system doesn't change unless externally influenced.

linear momentum: The product of mass times velocity, a vector.

magnetic field: The force on a moving positive test charge, per Coulomb, from magnets or moving charges.

magnitude: The size or length associated with a vector (vectors are made up of a direction and a magnitude).

mass: The property that makes matter resist being accelerated.

MKS system: The measurement system that uses meters, kilograms, and seconds.

moment of inertia: The property of matter that makes it resist rotational acceleration.

Newton: The MKS unit of force — one kilogram-meter per second^2.

normal force: The force a surface applies to an object, in a direction perpendicular to that surface.

ohm: The MKS unit of electrical resistance.

oscillate: Move or swing side to side regularly.

pascal: The MKS unit of pressure — 1 Newton per meter^2.

period: The time it takes for one complete cycle of a repeating event.

photon: A quantum of electromagnetic radiation. An elementary particle that is its own antiparticle.

polarize: Cause to vibrate in a definite pattern.

potential energy: An object's energy because of its position when a force is acting on it or its internal configuration.

power: The rate of change in a system's energy.

pressure: Force applied to an object divided by the area over which the force acts.

radians: The MKS unit of angle — 2π radians in a circle.

radiation: A physical mechanism that transports heat and energy as electromagnetic waves.

ray diagram: A drawing that shows how light rays travel through lenses or mirrors.

RC circuit: An electric circuit that contains both a resistor (or resistors) and a capacitor (or capacitors).

real image: An image that you can see on a physical screen.

refraction: The process that occurs when light crosses the boundary between two substances and bends.

resistance: The ratio of voltage to current for an element in an electric circuit.

resistor: An element in an electric circuit that resists the passage of current.

resultant: A vector sum.

rotational inertia: An object's moment of inertia.

scalar: A simple number (without a direction, which a vector has).

series: Elements arranged so the current that flows through one also flows through the other.

simple harmonic motion: Repetitive motion where the restoring force is proportional to the displacement.

solenoid: A wire coil.

special relativity: Albert Einstein's theory that gives explanations for the behavior of objects near the speed of light, such as time dilation and length contraction.

specific heat capacity: A material's heat capacity per kilogram.

standard pressure: One atmosphere — 1.0×10^5 Pascals.

standard volume: Defined as 22.4 liters.

static friction: Friction on an object that resists making it move.

thermal conductivity: A property of a substance showing how well or how poorly heat moves through it.

thermal expansion: The expansion of a material as it gets hotter.

thermodynamics: The section of physics covering heat and matter.

torque: A force around a turning point multiplied by the distance to that turning point.

uncertainty principle: A principle that says it's impossible to know an object's exact momentum and position.

vector: A mathematical construct that has both a magnitude and a direction.

velocity: The rate of change of an object's position, expressed as a vector whose magnitude is speed.

virtual image: An apparent image that isn't real — it can't appear on a physical screen.

volt: The MKS unit of electrostatic potential — one Joule per Coulomb.

weight: The force exerted on a mass by a gravitational field.

work: Force multiplied by the distance over which that force acts.

Index

BUSINESS, CAREERS & PERSONAL FINANCE

Grant Writing FOR DUMMIES
A Reference for the Rest of Us!
0-7645-5307-0

Home Buying FOR DUMMIES
A Reference for the Rest of Us!
0-7645-5331-3 *†

Also available:
- Accounting For Dummies †
  0-7645-5314-3
- Business Plans Kit For Dummies †
  0-7645-5365-8
- Cover Letters For Dummies
  0-7645-5224-4
- Frugal Living For Dummies
  0-7645-5403-4
- Leadership For Dummies
  0-7645-5176-0
- Managing For Dummies
  0-7645-1771-6

- Marketing For Dummies
  0-7645-5600-2
- Personal Finance For Dummies *
  0-7645-2590-5
- Project Management For Dummies
  0-7645-5283-X
- Resumes For Dummies †
  0-7645-5471-9
- Selling For Dummies
  0-7645-5363-1
- Small Business Kit For Dummies *†
  0-7645-5093-4

HOME & BUSINESS COMPUTER BASICS

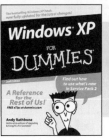

Windows XP FOR DUMMIES
A Reference for the Rest of Us!
0-7645-4074-2

Excel 2003 ALL-IN-ONE DESK REFERENCE FOR DUMMIES
9 BOOKS IN 1
0-7645-3758-X

Also available:
- ACT! 6 For Dummies
  0-7645-2645-6
- iLife '04 All-in-One Desk Reference
  For Dummies
  0-7645-7347-0
- iPAQ For Dummies
  0-7645-6769-1
- Mac OS X Panther Timesaving
  Techniques For Dummies
  0-7645-5812-9
- Macs For Dummies
  0-7645-5656-8

- Microsoft Money 2004 For Dummies
  0-7645-4195-1
- Office 2003 All-in-One Desk Reference
  For Dummies
  0-7645-3883-7
- Outlook 2003 For Dummies
  0-7645-3759-8
- PCs For Dummies
  0-7645-4074-2
- TiVo For Dummies
  0-7645-6923-6
- Upgrading and Fixing PCs For Dummies
  0-7645-1665-5
- Windows XP Timesaving Techniques
  For Dummies
  0-7645-3748-2

FOOD, HOME, GARDEN, HOBBIES, MUSIC & PETS

Feng Shui FOR DUMMIES
A Reference for the Rest of Us!
0-7645-5295-3

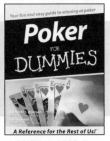

Poker FOR DUMMIES
A Reference for the Rest of Us!
0-7645-5232-5

Also available:
- Bass Guitar For Dummies
  0-7645-2487-9
- Diabetes Cookbook For Dummies
  0-7645-5230-9
- Gardening For Dummies *
  0-7645-5130-2
- Guitar For Dummies
  0-7645-5106-X
- Holiday Decorating For Dummies
  0-7645-2570-0
- Home Improvement All-in-One
  For Dummies
  0-7645-5680-0

- Knitting For Dummies
  0-7645-5395-X
- Piano For Dummies
  0-7645-5105-1
- Puppies For Dummies
  0-7645-5255-4
- Scrapbooking For Dummies
  0-7645-7208-3
- Senior Dogs For Dummies
  0-7645-5818-8
- Singing For Dummies
  0-7645-2475-5
- 30-Minute Meals For Dummies
  0-7645-2589-1

INTERNET & DIGITAL MEDIA

Digital Photography FOR DUMMIES
A Reference for the Rest of Us!
0-7645-1664-7

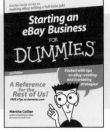

Starting an eBay Business FOR DUMMIES
A Reference for the Rest of Us!
0-7645-6924-4

Also available:
- 2005 Online Shopping Directory
  For Dummies
  0-7645-7495-7
- CD & DVD Recording For Dummies
  0-7645-5956-7
- eBay For Dummies
  0-7645-5654-1
- Fighting Spam For Dummies
  0-7645-5965-6
- Genealogy Online For Dummies
  0-7645-5964-8
- Google For Dummies
  0-7645-4420-9

- Home Recording For Musicians
  For Dummies
  0-7645-1634-5
- The Internet For Dummies
  0-7645-4173-0
- iPod & iTunes For Dummies
  0-7645-7772-7
- Preventing Identity Theft For Dummies
  0-7645-7336-5
- Pro Tools All-in-One Desk Reference
  For Dummies
  0-7645-5714-9
- Roxio Easy Media Creator For Dummies
  0-7645-7131-1

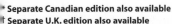

* **Separate Canadian edition also available**
† **Separate U.K. edition also available**

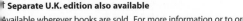

Available wherever books are sold. For more information or to order direct: U.S. customers visit www.dummies.com or call 1-877-762-2974.
U.K. customers visit www.wileyeurope.com or call 0800 243407. Canadian customers visit www.wiley.ca or call 1-800-567-4797.

SPORTS, FITNESS, PARENTING, RELIGION & SPIRITUALITY

0-7645-5146-9

0-7645-5418-2

Also available:
- Adoption For Dummies
  0-7645-5488-3
- Basketball For Dummies
  0-7645-5248-1
- The Bible For Dummies
  0-7645-5296-1
- Buddhism For Dummies
  0-7645-5359-3
- Catholicism For Dummies
  0-7645-5391-7
- Hockey For Dummies
  0-7645-5228-7

- Judaism For Dummies
  0-7645-5299-6
- Martial Arts For Dummies
  0-7645-5358-5
- Pilates For Dummies
  0-7645-5397-6
- Religion For Dummies
  0-7645-5264-3
- Teaching Kids to Read For Dummies
  0-7645-4043-2
- Weight Training For Dummies
  0-7645-5168-X
- Yoga For Dummies
  0-7645-5117-5

TRAVEL

0-7645-5438-7

0-7645-5453-0

Also available:
- Alaska For Dummies
  0-7645-1761-9
- Arizona For Dummies
  0-7645-6938-4
- Cancún and the Yucatán For Dummies
  0-7645-2437-2
- Cruise Vacations For Dummies
  0-7645-6941-4
- Europe For Dummies
  0-7645-5456-5
- Ireland For Dummies
  0-7645-5455-7

- Las Vegas For Dummies
  0-7645-5448-4
- London For Dummies
  0-7645-4277-X
- New York City For Dummies
  0-7645-6945-7
- Paris For Dummies
  0-7645-5494-8
- RV Vacations For Dummies
  0-7645-5443-3
- Walt Disney World & Orlando For Dummies
  0-7645-6943-0

GRAPHICS, DESIGN & WEB DEVELOPMENT

0-7645-4345-8

0-7645-5589-8

Also available:
- Adobe Acrobat 6 PDF For Dummies
  0-7645-3760-1
- Building a Web Site For Dummies
  0-7645-7144-3
- Dreamweaver MX 2004 For Dummies
  0-7645-4342-3
- FrontPage 2003 For Dummies
  0-7645-3882-9
- HTML 4 For Dummies
  0-7645-1995-6
- Illustrator CS For Dummies
  0-7645-4084-X

- Macromedia Flash MX 2004 For Dummies
  0-7645-4358-X
- Photoshop 7 All-in-One Desk
  Reference For Dummies
  0-7645-1667-1
- Photoshop CS Timesaving Techniques
  For Dummies
  0-7645-6782-9
- PHP 5 For Dummies
  0-7645-4166-8
- PowerPoint 2003 For Dummies
  0-7645-3908-6
- QuarkXPress 6 For Dummies
  0-7645-2593-X

NETWORKING, SECURITY, PROGRAMMING & DATABASES

0-7645-6852-3

0-7645-5784-X

Also available:
- A+ Certification For Dummies
  0-7645-4187-0
- Access 2003 All-in-One Desk
  Reference For Dummies
  0-7645-3988-4
- Beginning Programming For Dummies
  0-7645-4997-9
- C For Dummies
  0-7645-7068-4
- Firewalls For Dummies
  0-7645-4048-3
- Home Networking For Dummies
  0-7645-42796

- Network Security For Dummies
  0-7645-1679-5
- Networking For Dummies
  0-7645-1677-9
- TCP/IP For Dummies
  0-7645-1760-0
- VBA For Dummies
  0-7645-3989-2
- Wireless All In-One Desk Reference
  For Dummies
  0-7645-7496-5
- Wireless Home Networking For Dummies
  0-7645-3910-8